SpringerBriefs in Economics

More information about this series at http://www.springer.com/series/8876

Yanqing Jiang

Green Development in China

Models and Discussions

Springer

Yanqing Jiang
School of Economics and Finance
Shanghai International Studies University
Shanghai
China

ISSN 2191-5504 ISSN 2191-5512 (electronic)
SpringerBriefs in Economics
ISBN 978-981-10-0692-0 ISBN 978-981-10-0693-7 (eBook)
DOI 10.1007/978-981-10-0693-7

Library of Congress Control Number: 2016934943

Printed on acid-free paper

This Springer imprint is published by Springer Nature
The registered company is Springer Science+Business Media Singapore Pte Ltd.

Acknowledgements

The research contained in this book was carried out partly under a key research project of Shanghai International Studies University started in the year 2015. The author thus thanks the funding support for the project from Shanghai International Studies University.

Contents

Chapter 1
Introduction

A growing concern China and most developing countries face is that the pattern of economic development may not be sustainable because it tends to lead to irreversible depletion of natural resources and deterioration of the ecological environment. The relationship between economic development, pollution, and the environmental quality is thus an important issue that deserves careful study. Environmental considerations play an important role in shaping China's development strategies. Green development of China creates strong pressures for continuous transforming, upgrading and restructuring of the Chinese economy. The present book explores how China's economic development is related to pollution and the environmental quality, and how considerations about issues associated with the eco-environmental system may affect China's development strategies. The book presents to the reader facts, thoughts, models, empirical results and discussions that shed light on the related issues.

This introduction chapter introduces the topics and main contents of the subsequent chapters. Chapter 2 presents a non-technical overview of China's development from an environmental perspective.

In Chap. 3 we follow the concept of comprehensive wealth and present some preliminary empirical results concerning China's potential sustainable growth. The core idea is an extended concept of wealth, namely, the concept of 'comprehensive wealth'. We build our analysis of Chap. 3 on the very essential idea of comprehensive wealth. The idea takes account of both levels and changes in natural capital, human capital, reproducible capital as well as health capital and knowledge capital in its theoretical framework. The different types of capital jointly generate a measure of comprehensive investment.

In Chap. 4 we provide a preliminary discussion of economic development and the Environmental Kuznets Curve (EKC) in the case of China. Chapter 4 serves as a precursor study to later chapters. Among other findings, empirical results in Chap. 4 show that with regional population controlled for in our regressions, as regional

Y. Jiang, *Green Development in China*,
SpringerBriefs in Economics, DOI 10.1007/978-981-10-0693-7_1

income or regional per capita income grows larger, pollution emission and emission intensity first increases and then declines. This result lends support to the inverted U-shaped EKC hypothesis. Moreover, empirical results in Chap. 4 also show that physical capital accumulation does not have a significant effect on pollution emission, but a higher level of human capital intensity helps contribute to a lower level of per capita pollution emission.

In Chap. 5 we approach the environmental issue from a somewhat different angle. We present an empirical discussion of the relationship between economic growth and the environmental input, based on related empirical evidence from the Chinese regions. The study in Chap. 5 explores whether the use of environmental resources taken as a factor of production contributes crucially to regional economic growth in the case of the Chinese regions. Empirical results in Chap. 5 show that for the Chinese regions, a significant part of output growth and productivity growth could actually be attributed to the increasing use of environmental inputs.

In Chap. 6 we focus on the relationship between foreign trade and sustainable development in the case of the Chinese regions. Foreign trade may affect pollution outcomes and the environment via various different channels. In Chap. 6, based on some theoretical framework, we will empirically examine the effects of trade openness and the level of development on pollution emission across the Chinese provinces. Our empirical results in Chap. 6 suggest that when the effects of per capita income, population, and per capita physical capital and human capital stocks are netted out, openness to foreign trade exerts a significant positive partial effect on per capita pollution emission. Our empirical results here also support the EKC hypothesis by showing that as income or TFP increases pollution emission and emission intensity both first rise and then decline.

Continuing from our discussions in the Chap. 6, we move on to investigate the relationship between foreign direct investment (FDI) and sustainable development in the case of the Chinese regions in Chap. 7. Here we are interested in uncovering the environmental impacts of China's processes of economic development and opening up to FDI. Specifically, we empirically examine the potential effects of FDI and the level of economic development on pollution emissions across the Chinese regions. We find that when the effects of income, population, physical and human capital are accounted for, regional FDI tends to have a significant positive effect on regional pollution emission. Like Chap. 6, the study in Chap. 7 also provides support to the EKC framework.

In earlier chapters we take the Chinese regions as isolated geographical units and we have not considered the potential spatial interactive behavior among the regions regarding pollution emission and pollution abatement efforts. In Chap. 8 we move on to consider the possibility of spatial strategic interaction in pollution abatement efforts across the Chinese regions. Based on a theoretical framework we reach our empirical reaction function, where the sign and magnitude of the reaction slope are of our primary interest. Our empirical results in Chap. 8 show that the Chinese provincial governments tend to behave strategically in deciding on the levels of their environmental protection efforts. The significantly positive estimates of the reaction slope suggest that environmental protection efforts of the provincial governments are 'strategic complements'.

Finally, Chap. 9 looks into the future and presents a brief discussion about some future objectives of China's long march in green development.

Chapter 2
Development and the Environment in China: An Overview

China, as the biggest developing country in the world, has seen rapid increases in income levels for many decades. However, in recent years the average annual growth speed as measured by the GDP index has decreased to about 7 %. The industrial sectors have also experienced huge transformation. The primary and tertiary industries are both growing steadily at a 4 and 8 % annual rate respectively while the growth rate of the secondary industry has dropped markedly during 2010–2014. The secondary industry in China usually relies heavily on natural resources and cheap labor. However, the comparative advantage of neither of these resources will last forever. In the past decades, economic development in China relied on high rates of investment, huge energy consumption and excessive environmental pollution. The processes of urbanization, industrialization and infrastructure construction in China need continuous investments. To some extent, the traditional development pattern in China has reached its limit. Extensive environmental pollution is now posing a serious threat to human health and further economic development.

China's industrialization and development has consumed so much of natural resources and created a great deal of pollution. Smoke and acid rain caused by the coal industry are becoming more and more of a serious concern, especially in Central China. Water pollution in industrial cities is endangering not only the animals and plants, but also the drinking water of city residents. Statistics show that two thirds of China's rivers and more than ten million hectares of farmlands are polluted. Taking SO_2 and smoke as representatives of pollutant gases, we can see neither of them has experienced a great decrease during 2010–2014. The emission of wastewater even raised a lot in the recent five years. However, according to the statistics, the Chinese government has invested great money in industrial pollutant governance and environmental protection. However, the practical results of the government's investment in pollutant governance are far from being satisfactory, and the resulting pollution constitutes a hindrance to the sustainable development of the Chinese economy.

Y. Jiang, *Green Development in China*,
SpringerBriefs in Economics, DOI 10.1007/978-981-10-0693-7_2

According to the Environmental Kuznets Curve (EKC) model, the relationship between environmental pollution and economic development can be traced by an inverted- U curve, which means that environment gets worse with economic improvement until a certain point, and then becomes better along with increases in the residents' average income. The EKC analyzes such a relationship from the perspectives of the scale and structure effects of the economy, the function of technology, as well as the demand for environmental regulations. Although the EKC may not suit all regions or all types of pollutants, it still roughly manifests a general trend of changes in the environment quality and informs the choice in policy-making. Useful conclusions concerning the EKC can be drawn from our empirical analysis in later chapters of this book. Among other findings, empirical results in this book show that with regional population controlled for in our regressions, as regional income or regional per capita income grows larger, pollution emission and emission intensity first increases and then declines. However, China has a long way to go in improving its environmental quality and achieving a process of green development. The degradation of China's natural environment will not eventually be curbed if long-term measures are not implemented in time.

Since China's economic reform and opening up, rapid economic growth has induced massive consumption of energy and natural resources. This led to heavy emissions of industrial pollution and the resulting exacerbated damage to the natural environment. Early economic reform was implemented at a heavy cost in China's ecology and natural environment. The mid-1990s was a period in which China began to realize the severity of its environmental problem and came to follow a sustainable development strategy. Subsequently, China made efforts in achieving improvement in energy utilization efficiency so as to narrow its natural deficit. The environmental loss measured as part of GDP dropped considerably. As China entered the new millennium, the concept of green development became an integral part of the notion of 'science based development'. In the new century, China sees an ecological surplus. There has been a significant decline in GDP losses through natural disasters while there has been a substantial increase in green investment.

The Five-Year Plan is an important dividing scheme for the planning decisions made by the Chinese government. The period of the First Five-Year Plan was an important period in which China undertook nearly two hundred major industrial projects, with the energy industry given the top priority. The major industrial projects consumed huge amounts of China's energy, resources and raw materials. The growth rate of energy consumption in that period by far exceeded that of economic growth. Economic growth during the time of the 'Great Leap Forward' was mostly due to the development of ecologically costly industries throughout China. Growth was achieved at a huge cost of in environmental and in human resources. Subsequently, to remedy the industrial chaos during the Great Leap Forward, China placed renewed emphasis on central planning control over material deployment and investment in production to reduce energy waste. During the period following the Great Leap Forward, the fueled by high consumption of resources, the Chinese economy grew at a quite high pace. The growth was accompanied by high levels of pollution emissions. With the development of its energy industry, China underwent the transformation from an impoverished, self-sufficient country to one that was a major producer of resources. But China followed a resource-intensive and energy-intensive path of industrialization.

The key goals of the Five-Year Plans were to promote China's social progress and economic development. The Plans, however, underwent a shift of emphasis from economic plans, to strategic plans, and then to comprehensive development plans over time. During the process, the Plans have shifted the weight from promoting a 'black' (polluting) economy to promoting a green economy. During the first to fifth Five-Year Plans, the main objectives were to develop the key industries and to promote the process of basic industrialization. Therefore, the early Five-Year Plans were characterized by their emphasis on heavy industry. The First Five-Year Plan, as mentioned already in the preceding paragraph, established the building up of the heavy industry as a priority. The Second Five-Year Plan placed the focus on the development of the steel industry while the following two plans, the Third and Fourth Five-Year Plans, established the strategic importance of the steel industry as the foundation for military preparedness. The Fifth Five-Year Plan aimed to initiate big progress in the manufacturing sector. The first five Five-Year Plans thus set the basic tone for China's 'black' industrialization. The 'black' industrialization was characterized by high levels of resources, energy consumption and pollution emissions.

Along with the initiation of China's reform and opening up, the Five-Year Plans also underwent consideration transformation. They gradually shifted their emphasis from solely economic planning to social development planning. The Plans progressively abandoned the path of the 'black' economy. Unlike the old approach, the Sixth Five-Year Plan placed an emphasis on steady and healthy development of the national economy. The Seventh Five-Year Plan aimed at promoting production through intensive utilization of capital and resources. The Eighth Five-Year Plan laid an emphasis on development and conservation through efficient energy use. The Ninth Five-Year Plan introduced two changes. That is, the change of the economic system from a traditional planned economy to a socialist market economy, and the change of the economic growth mode from an extensive mode to an intensive mode. The Ninth Five-Year Plan period signaled that China began to experience a transition from a period of 'black' development to one of green development. The Tenth Five-Year Plan saw even greater progress toward green development. In this period, objectives in ecological and environmental protection for the first time became the main objectives of a Five-Year Plan.

The Eleventh Five-Year Plan was the first Plan being designed under the guidance of the central government's 'scientific outlook for development'. In this Plan, 30 % of its objectives were associated with green development, with particular prominence given to energy conservation and environmental protection. The Twelfth Five-Year Plan was China's first full-fledged green development plan. This Plan contained both qualitative and quantitative objectives for wide variety of ecological and environmental targets. The Plan proposed to establish a resource-saving, environment-friendly society and to improve the level of ecological civilization. The Plan also aimed to save energy consumption, improve energy utilization efficiency, and adjust the structure of energy uses. One section of the Twelfth Five-Year Plan was devoted to green development and set green development as a principle for ecological construction.

During the decade of the Eleventh and Twelfth Five-Year Plans periods, China's environmental indicators were better than those in the periods of previous Plans. For example, the trend of reduction in arable land was effectively controlled. Water consumption per unit of industrial added value declined markedly. Efforts made toward environmental and ecological protection began to show their effects. Deterioration in the ecological environment was initially curbed. Conservation areas were effectively protected and forest coverage was increased substantially. The proportion of green development indicators increased considerably in the Twelfth Five-Year Plan period. In addition, the Twelfth Five-Year Plan placed an emphasis on comprehensive and harmonious sustainable development of the economic system and social system. This Plan was a green plan in the sense that it aimed achieve dynamic growth through green development, wealth and welfare.

We end this chapter by presenting some detailed facts regarding the improvement as measured by the changes in major eco-environmental indicators during the past ten years of the Eleventh and Twelfth Five-Year Plans periods. The indicators for sulfur dioxide emissions and COD (chemical oxygen demand) emissions were increased by about 12 and 14 % points respectively during the Eleventh Five-Year Plan periods (2005–2010). The indicator for industrial solid waste treatment rate rose by about 12 % points during 2005–2010 and by about 5 % points during Twelfth Five-Year Plan period (2010–2015). The indicator for water consumption per unit of industrial value added rose by about 43 % points during 2005–2010 but dropped by about 10 % points during 2010–2015. The indicators for urban sewage treatment rate and urban life garbage treatment rate were raised by about 21 and 17 % points respectively during the Eleventh Five-Year Plan periods and by about 16 and 15 % points during the Twelfth Five-Year Plan period. The indicator for efficient utilization of agricultural irrigation water was increased by about 0.05 % points during 2005–2010 and then by about 6 % points during 2010–2015. The indicator for forest coverage rose by about 2 % points during 2005–2010 and then by about 6 % points during 2010–2015. The indicator for the proportion of state-controlled sections of seven major river systems better than Class III rose by about 18 % points during 2005–2010. Also during 2005–2010, the indicator for the proportion of cities with air quality standard above Class II rose by about 23 % points and that for natural wetland protection rose by about 5 % points. In the Eleventh Five-Year Plan period, the growth rates of two representative pollutant emissions, namely, industrial chemical oxygen demand emissions and industrial sulfur dioxide emissions were both negative, being −1.66 and −2.60 %, respectively.

Chapter 3
A Broadened Concept of Wealth and Sustainable Development

3.1 Introduction

A growing concern that China and most other developing countries face is that the patterns of their economic growth and development cannot be sustained because they tend to result in irreversible depletion of natural resources and deterioration of the ecological environment. It is mostly such concerns that have helped spawn an ever-increasing body of relevant literature on sustainable growth and development. However, There still remains a considerable challenge for scholars and policy makers to arrive at a theoretically consistent and empirically feasible criterion for evaluating policies that target sustainability in economic growth and development—especially under the current fast-growing climate of globalization.

To evaluate the effectiveness of a development policy—in term of environmental sustainability, we need to at least incorporate environmental factors into the national income system—mostly by expanding income measures—to make it account not only for the returns to labor and capital, but also for changes in stocks of natural sources. With the natural environment being taken into consideration, sustainability means enabling future generations to achieve a level of well being at least as high as that of the current generation, where traditional income accounts have failed to provide a useful tool for measuring and evaluating human well being associated with natural resources and the natural environment. In this vein, the World Bank (2006) used the 'adjusted net savings', which was calculated as national saving minus the values of environmental damage and depletion of natural capital, as the criterion for sustainability. In this sense, sustainable development is present if and only if the adjusted net savings are positive.

However, an obvious shortcoming of the use of adjusted net savings as a criterion for sustainability is that this measure pertains to the change in wealth, but not directly to the level of wealth. It is better to focus directly on the level of wealth if what we are interested in is the long-run sustainability in the growth of human welfare. In other words, if sustainability corresponds to the increase in the potential

Y. Jiang, *Green Development in China*,
SpringerBriefs in Economics, DOI 10.1007/978-981-10-0693-7_3

welfare of a typical individual, what we should focus on is then the increase in per capita wealth, measurement of which is based on the level of wealth as well as on changes in it. Another shortcoming of adjusted net savings is that this measurement does not take account of changes in factors such as human capital, technology and the life span when measuring the changes in 'wealth'.

Seeing the shortcomings, the World Bank (2011) improved on its 2006s approach by incorporating the values of some important factors as components of a comprehensive measure of wealth. These factors include the values of human, natural and reproducible capital. However, to date, the World Bank and others have not successfully considered the values of many important forms of intangible capital when forming their measures of wealth. It is therefore necessary to come up with a more comprehensive measure of welfare or wealth in order to better assess the likelihood of real sustainability in economic development.

In this chapter, we basically follow the idea of Arrow et al. (2012) and present some preliminary empirical results concerning China's potential sustainable growth. The core of Arrow et al. (2012)'s idea is an extended concept of wealth, namely, the concept of 'comprehensive wealth'. We therefore base our analysis of this chapter on the very essential idea of comprehensive wealth. The idea takes account of both levels and changes in natural capital, human capital, reproducible capital as well as health capital and knowledge capital in its theoretical framework. The different types of capital jointly generate a measure of comprehensive investment. The organization of this chapter goes as follows. Section 3.2 introduces the concept of comprehensive wealth and discusses sustainable development from the perspective of comprehensive wealth. Section 3.3 presents some specific results from an empirical analysis of China's potential sustainable growth. Finally, Sect. 3.4 concludes this chapter.

3.2 Sustainable Development in Terms of Comprehensive Wealth

The general idea of sustainable development put forward by Arrow et al. (2012) is that economic development is considered sustainable at any given point in time if and only if at that moment the time derivative of the so-called intergenerational welfare is nonnegative. The intergenerational welfare is in turn calculated as the discounted flow of the instantaneous felicities (utilities) of current and future generations. Without loss of generality, we can write intergenerational welfare $V(t)$ as the following, assuming for the moment population is constant

$$V(t) = \int_{t}^{\infty} [U(\mathbf{C}(s))e^{-\delta(s-t)}ds] \tag{3.1}$$

where $U(\mathbf{C}(s))$ is the instantaneous social felicity (utility) at time s with $\mathbf{C}$ being a vector of consumptions, and δ is the felicity discount rate. The economic forecast at time t is then captured by the pair of vectors $\{\mathbf{C}(s), \mathbf{K}(s)\}$ for $s \geq t$. We can define prospective sustainable development at time t as being equivalent to saying $dV/dt \geq 0$.

We can then see that even without explicitly writing down the dynamic evolution of the economy in development, the future of the economy depends on the evolving characteristics of the capital stocks, together with technology, people's preferences, as well as factors such as the institutions. By continuing from moment to moment, the entire future course of the evolution of capital stocks can be determined. It can therefore be shown that intergenerational welfare can eventually be expressed as

$$V(t) = V(\mathbf{K}(t), t) \tag{3.2}$$

where intergenerational welfare at a given point in time, $V(t)$, depends directly on time because of the effects of time-varying factors are treated as exogenous such as changes in technology, the terms of trade, unexpected growth in population and unexpected shifts in institutions. In this sense, the time t can be conveniently regarded as an additional capital asset. Using (3.2), sustainable development can be written as

$$dV(t)/dt = \partial V(t)/\partial t + \sum_i [(\partial V(t)/\partial K_i(t))(dK_i(t)/dt)] \geq 0 \tag{3.3}$$

where $K_i(t)$ denotes the ith capital asset at time t.

The comprehensive wealth of an economy can then be defined as

$$W(t) = r(t)t + \sum_i p_i(t)K_i(t) \tag{3.4}$$

where $p_i(t) \equiv \partial V(t)/\partial K_i(t)$ is the shadow price of the ith capital asset at time t, and $r(t) \equiv \partial V/\partial t$ is the shadow price of time at t. It can be shown that a small perturbation to an economy increases (decreases) intergenerational welfare if and only if it increases (decreases) comprehensive wealth, holding the shadow prices constant (see Arrow et al. 2012 for a proof). Further we can define comprehensive investment as

$$\Delta V(t) = r(t)\Delta t + \sum_i p_i(t)I_i(t)\Delta t \tag{3.5}$$

where $I_i(t) \equiv \Delta K_i(t)/\Delta t$. Therefore, a small perturbation to an economy increases (decreases) intergenerational welfare at time t if and only if the shadow value of comprehensive investment at t associated with the perturbation is positive (negative). Integrating (3.5) above leads to

$$V(T) - V(0) = \int_0^T r(s)ds + \sum_i [p_i(T)K_i(T) - p_i(0)K_i(0)] - \int_0^T [\sum_i (dp_i(s)/ds)K_i(s)]ds \quad (3.6)$$

which reveals that in detecting an increase in intergenerational welfare between two dates, the accrued capital gains on the assets over the interval should be deducted from the difference in wealth between the dates.

Of course the basic framework above can be extended in many ways. For example, a change in technology may involve investment in research and development (R&D) while expenditure in R&D is a part of comprehensive investment. Assume $Y(t) = A(t)F(\mathrm{K}(t))$, where $F(\bullet)$ is a constant returns to scale function, and $A(t)$ is taken to be total factor productivity (TFP) at time t, which may serve as an index of knowledge or/and the economy's institutions. On a balanced growth path, it can be shown that

$$\partial V/\partial t = \gamma q_A(t)A(t)/[\sum_i p_i(t)K_i(t) \quad (3.7)$$

where γ is the growth rate of TFP and $q_A(t)$ is the shadow price of TFP. Empirically, we need merely to add TFP growth to comprehensive investment Eq. (3.5) as we can show that when the rate of national saving is small, the factor $q_A(t)A(t)/[\sum_i p_i(t)K_i(t)$ is approximately equal to unity.

It should be noted that if otherwise growth in TFP is taken to be exogenous (say, TFP accrues through learning by doing, rather than through R&D), it is then simply reflected in $r(t) \equiv \partial V/\partial t$ [or the first term of (3.5)].

Population growth induces complications to our analysis. When the size of the population is not fixed, the criterion for sustainable development should be non-decreasing comprehensive wealth in per capita terms. Specifically, it can be shown that development is sustained at time t if and only if per capita comprehensive wealth is non-declining at t when valued at constant shadow prices.

3.3 Some Empirical Issues and Results

Having introduced the concepts and stated the propositions in the preceding section, we now turn to the empirical aspects of the issue in this section. The main point in carrying out empirical work lies in the proper measuring of the levels of and changes in the stocks of various types of capital. Further, we may need to aggregate those levels and changes in order to be able to estimate comprehensive investment and comprehensive wealth, which requires using shadow prices in constructing the stocks of various types of capital.

Two prominent publications, the World Bank (2006, 2011), made an effort in estimating wealth and its components in 120 countries over time (specifically, in the years 1995, 2000 and 2005). Comprehensive wealth was defined as the present value of the flow of aggregate future consumption. Growth rates in consumption for the foreseeable future are first predicted, and then discounted values of the consumption stream are computed in order to estimating comprehensive wealth. Shadow values of reproducible capital and environmental capital are estimated and subtracted from comprehensive wealth so as to reach a value referred to as intangible capital, which includes human capital, public knowledge, institution, etc. Environmental capital was taken to include energy, mineral, and forest resources as well as agricultural, urban and pasture lands. By following such a process of measuring and estimating, the World Bank (2006, 2011) concluded that in poor economies the shadow value (the sum of various amounts multiplied by corresponding shadow prices) of environmental capital took up about one quarter of the total amount of comprehensive wealth, and that the proportion of intangible capital in comprehensive wealth exceeded one half (about 55 %) of the total amount of the latter.

Compared with the World Bank (2006, 2011), Arrow et al. (2012) followed a different approach in calculating its levels and changes of the various components of comprehensive wealth. The latter argues that assuming the availability of a good forecast of the future stream of consumption amounts to saying that we know how sustainable the economy will work out, which is exactly what we are trying to estimate. Specifically, Arrow et al. (2012) constructed estimates of the levels of the stocks of the various types of capital and summed them up to form comprehensive wealth, where no form of comprehensive wealth is calculated simply as a residual.

As to the environmental capital, to reach values of net investment in it, changes in stocks of various environmental resources as well as the individual shadow prices are needed. For nonrenewable resources such as metal deposits, the change in the stock is simply the negative of the amount depleted (extracted) during the given period. The rental value can be taken to correspond to the shadow price of that resource if we neglect externalities arising from the use of the resource. For renewable resources such as forests, net investment equals the increase in the forests from natural growth and replanting, subtracted from which the amount depleted. The shadow price is again taken to be the rental value (which is the price less the cost of cutting).

Moreover, the broad concept of environmental capital necessarily encompasses a component that is associated with environmental externalities due to, say, air and water pollution caused by the extraction of natural resources or the creation of manufactured wealth. These externalities constitute a drawdown of wealth. Arrow et al. (2012) focused on only one type of such environmental externality, which was climate related damages. Unlike with other forms of natural capital, Arrow et al. (2012) did not attempt to estimate the total level of the environmental asset (the global atmosphere), but only considered the change in this environmental asset.

To apply the estimation procedure, data on global emissions of greenhouse gases are needed, so are data on the damage to each of the individual countries associated with these emissions. Arrow et al. (2012) obtained data on global emissions of carbon dioxide from fossil fuel consumption and manufacturing from the 2007 World Development Indicators data. Emissions of carbon dioxide from deforestation were calculated by employing data from the Food and Agriculture Organization (2006) and Houghton (2005). Their measure of total carbon dioxide emissions over the 1995–2000 period was 35.5 billion tons and likely underestimated the true emissions over that period since some key sectors were not included, and they also used conservative estimates of carbon emissions from deforestation.

Following the method by Tol (2009), which was a meta-analysis of 232 studies concerning the social cost of carbon, Arrow et al. (2012) assumed that the damage from global emissions is $50 per ton carbon, for which alternative values were also considered in a sensitivity analysis. Combining the damages per ton and emissions data, Arrow et al. (2012) ended up with total damages of $1840 billion for the five-year period of 1995–2000. Applying estimates from Nordhaus and Boyer (2000), Arrow et al. (2012) assigned the total damages to the five countries under study. They apportioned the damages to each country as follows: the U.S. bears nine percent of global loss, India bears five percent, Brazil bears two percent, China and Venezuela both bear less than one percent.

Generally, taking the wide variety of capital forms into account, Arrow et al. (2012) estimated high rates of overall and per capita comprehensive wealth growth for China. However, it was shown in their study that the high growth rates of comprehensive wealth were mostly due to China's significant increase in reproducible capital during the period of 1995–2000. The final conclusion of Arrow et al. (2012) for China was that China was currently meeting the sustainability criterion. Based on their analyses, two other countries under study, the United States and India, were also meeting the sustainability criterion. In those two countries investments in human capital were thought to be a very important contributor to increases in per capita wealth while in contrast China's investments in reproducible capital were shown to be the main contributor.

Sustainable development literature had its roots in the concern about 'greening' the national accounts. In this sense the work of Arrow et al. (2012) was a welcome addition to the literature that shows how expanding the asset boundary beyond produced and natural goods could increase our measures of income, saving and wealth enormously (Hamilton 2012; see also Solow 2012). Arrow et al. (2012) also provided a formal definition for comprehensive wealth, which stated that 'an economy's comprehensive wealth is the shadow value of all its capital assets. Based on this definition. As mentioned already, Arrow et al. (2012) concluded that development is sustained at time t if and only if, when valued at constant shadow prices, comprehensive wealth per capita is non-decreasing at time t.

However, based on a different theoretical framework from that of Arrow et al. (2012), one of our own prior studies, Jiang (2015), reached a major conclusion that deviated substantively from that of Arrow et al. (2012). We sketch our analysis in Jiang (2015) in the following paragraphs.

In Jiang (2015), we started with a discussion of a green Solow growth model by following the basic procedures of Brock and Taylor (2004) and Copeland and

Taylor (1994). Based on the theoretical green growth model we then moved on to the empirical discussion starting with a Hicks-neutral Cobb-Douglas aggregate production function $Y = AK^{\alpha}L^{1-\alpha}$, where obviously, Y is total output, K is capital stock, L is labor (or equivalently population or the number of works), and A stands for the level of technology (knowledge). Per capita output can thus be written as

$$y = A^{\frac{1}{1-\alpha}}\left(\frac{k}{y}\right)^{\frac{\alpha}{1-\alpha}} \tag{3.8}$$

where y and k are per worker output and per worker capital stock, $y \equiv Y/L$ and $k \equiv K/L$. We define growth of y due to growth of A as 'intensive' growth, and growth of y due to an increase of k/y as 'extensive' growth. It should be noted that the decomposition of per capita output in Eq. (3.8) is performed with respect to the capital-output ratio, rather than the usual capital-labor ratio for two reasons. As Klenow and Rodriguez (1997) and Hall and Jones (1999) rightly pointed out, first, along a balanced growth path, the capital-output ratio is proportional to the investment rate so as to render this decomposition a natural interpretation. Second, if we had performed the decomposition in with respect to the capital-labor ratio, we would have improperly attributed part of the increase in per capita output that would be fundamentally due to an increase in TFP to capital accumulation.

Inspired by the idea we gain from the green growth model, we examine the relationship between per worker output and per worker pollution emission across the Chinese provinces.[1] With data available from various official sources and proxying for the total regional pollution emission of a Chinese province coarsely by the total volume of regional industrial waste gas emission, results from various regressions showed that across all the Chinese regions a one percent increase in the level of real regional per capita output would on average lead to a 0.65 % increase in the level of regional per capita pollution emission.[2] Results also showed that an increase of one percentage point in the annual growth of real regional per capita output would on average lead to an increase of roughly 1.5 % points in the annual growth of regional per capita pollution emission.[3]

---

[1]Our sample in Jiang (2015) was 28 Chinese province-level regions over the period 1997–2011. These regions include provinces, ethnic minority autonomous regions, and province-level municipalities, but for convenience's sake we call all of them 'provinces'. Owing to incomplete data, Tibet, Chongqing and Hainan are not included in our sample.

[2]Series of the total volume of industrial waste gas emission, series of nominal Gross Regional Product (GRP), GRP indices, and numbers of total employed people (1997–2011) for each province are directly available from the Chinese Statistical Yearbooks (1997–2012), so that values of real GRP (1997–2011) for each province can be calculated. Real per worker output is calculated as real GRP divided by the number of total employed people for each province. Per worker pollution emission is calculated as the total volume of industrial waste gas emission divided by the number of total employed people for each province.

[3]The Chinese Statistical Yearbooks, however, do not directly record data on physical capital stocks for the Chinese regions. To obtain annual data on real provincial capital stocks, we follow the basic procedure of Zhang et al. (2007) and Zhang (2008), and use a perpetual inventory method (PIM) to

We were more interested in seeing whether the growth pattern (i.e. extensive/intensive growth) had an effect on pollution emission. Specifically, we want to see whether it is the variation in total factor productivity (TFP) or that in the degree of 'extensive growth' (as shown by the capital-output ratio k/y) that accounts for the larger share in the variation of pollution emission. Our regression results showed that across regions a one percent increase in the capital-output ratio k/y would on average lead to a roughly 0.3 % increase in the level of per capita pollution emission, and that for any given region, a one percent increase of the capital-output ratio would result a 1.64 % increase per capita pollution emission. To get a closer look into the issue, we re-ran the regressions but now included the full set of time intercepts to capture potential effects of growing TFP on pollution emission. Once we controlled for the year dummies, it turned out that none of the regressions would produce significant estimates for the coefficient on (the log of) the capital-output ratio. The result suggested that it was not the variation in the degree of 'extensive' growth, but the variation in TFP that played a crucial role in explaining the variation in pollution emission.

The central focus of Jiang (2015) was to investigate the relationship between TFP growth and pollution in China. In terms of an aggregate production function, output growth is driven by the collective contributions of its different sources. The different sources are the individual factors of production plus the so-called TFP. Usually, the concept of TFP growth was created to measure that part of output growth not accounted for by the growth of the different factors of production. In this sense growth accounting is still a useful tool in growth empirics, but much more shape and substance should be provided for the amorphous term TFP (Easterly and Levine 2001). In response to this need, Jiang (2015) made an effort to provide more shape and substance to the amorphous TFP, or the Solow residual, mostly by tracing a large chunk of it to environmental resources. In an unregulated market environmental resources are unpaid factors of production as the cost of pollution is not internalized. Neglecting to take the pollution costs of economic growth into account would mostly lead to biased conclusions (Brock 1973). The use of environmental resources as production factors can be addressed by including pollution emissions as an additional production input in the production function (Tzouvelekas et al. 2006).[4]

(Footnote 3 continued)

construct physical capital stock data for the Chinese provinces—where we specifically assume that the annual depreciation rates of physical capital are uniformly 9.6 % for all the provinces throughout our sample period. Once the values of real provincial capital stocks are obtained, it is then straightforward to calculate the corresponding values of real provincial per worker capital stocks.

[4]In this context, Tzouvelekas et al. (2006) refered to a few examples of the literature in which the production function was specified to include the flow of pollution emissions as an input and sometimes, productivity enhancing environmental quality as a stock variable: Brock (1973), Becker (1982), Tahvonen and Kuluvainen (1993), Bovenberg and Smulders (1995), Smulders and Gradus (1996), Mohtadi (1996), Xepapadeas (2005), Brock and Taylor (2005), and Considine and Larson (2006).

Therefore Jiang (2015) attempted to explore the linkage between TFP growth and pollution emissions across the Chinese provinces. The study aimed to investigate whether the use of the environment as a production input contributed crucially to output growth in the case of the Chinese provinces. Based on a regression framework of the form

$$\Delta \ln y_{it} = \beta_0 + \beta_1 \Delta \ln m_{it} + \beta_2 \Delta \ln k_{it} + c_i + u_{it} \tag{3.9}$$

where c_i is the time-constant province heterogeneity, u_{it} is the zero-mean idiosyncratic error term, empirical results of Jiang (2015) showed that FE regression produced an significant estimate of β_2 which was 0.470, with a 95 % confidence interval of [0.423, 0.518], and an significant estimate of β_1 which was 0.070, with a 95 % confidence interval of [0.050, 0.089].[5] All these results showed that showed that for the Chinese regions a significant part of output growth that was traditionally attributed to TFP growth could indeed be rightly attributed to the increasing use of environmental inputs. In other words, TFP of the Chinese provinces traditionally estimated in most of the previous literature should not be taken as totally 'green'. By running the regressions above, we effectively chipped away from the traditional TFP a chunk of 'environmental resource' that could be taken as an additional factor of production that contributed to the production of output. Our study thus suggested that in the process of China's economic growth (at least over the chosen sample period), even if we partialed out the effect of capital accumulation, output growth was still being achieved at the cost of increasing pollution emissions.

Further, a closer look at the parameters underlying the beta's suggested that long-run green (that is, environmentally sustainable) growth was unattainable in the case of the Chinese regions under current conditions of pollution abatement efforts.[6] Greater efforts from the Chinese central and local governments were needed to tackle problems related to pollution emission and environmental degradation in order to achieve China's green sustainable growth in the decades to come.

[5]Alternative regressions generated results similar to those of the FE regression. The RE GLS regression generated an estimated β_2 that was 0.460, with a 95 % confidence interval of [0.414, 0.506], and an estimated β_1 that was 0.069, with a 95 % confidence interval of [0.050, 0.089]. The plain pooled cross section OLS regression produced an estimated β_2 that was 0.439, with a 95 % confidence interval of [0.392, 0.485], and an estimated β_1 that was 0.069, with a 95 % confidence interval of [0.048, 0.090]. Jiang (2015) also found that the inclusion of the year dummies did not alter the regression results in any important ways.

[6]Given our assumptions made on China's green growth and results obtained from our regressions, we did not predict long-run environmentally sustainable growth for China. Moreover, as a byproduct, we were able to estimate the structural parameter α in the aggregate production function. We ended up with an estimated value of α that was around 0.50, in the very vicinity of its empirically accepted values for the case of China (Zheng et al. 2009; Brandt and Zhu 2010; Jiang 2011, 2012). Normally the structural parameter α varies across nations. For example, according to Zheng et al. (2009), for the United States, α was 0.30. (See also Congressional Budget Office 2001), for the EU, it was about 0.40 (See also Musso and Westermann 2005), and for China, it was around 0.50 (see also Chow and Li 2002 and Chow 2008). Therefore, this piece of by-product result lent support to our main results here in this study.

3.4 Concluding Remarks

Undoubtedly China has achieved spectacular economic growth and transition in the past three decades or so. Nevertheless, various environmental problems have arisen to become more and more of serious concerns in the country. Using GDP as the most prominent index for measuring economic performance does not hide the fact that the rapid growth of China's economic output has been achieved at the cost of severe environmental deterioration and resource degradation. Since the Tenth Five-Year Plan period (2001–2005), the Chinese government has been making great efforts to tackle the problems of environmental deterioration and resource degradation. The Eleventh Five-Year Plan period (2006–2010) saw a significant improvement in China's environmental performance over that of the previous (the Tenth) Five-Year Plan period.

Although the environmental targets set by the Eleventh Five-Year Plan turned out to have led to remarkable environmental achievements compared with the preceding Plan, yet those environmental improvements under the Eleventh Five-Year Plan were accomplished ignoring the fundamental fact that economic growth in China at that time relied too heavily on industrial investment but not sufficiently on developments in the service sector, and thus was not efficiently conducive to environmentally sustainable development. Concerning environmental protection, the Twelfth Five-Year Plan (2011–2015), in a broad sense, aimed to continue the strategies successfully implemented under the previous Plan. By aiming at strengthening environmental management, the Twelfth Five-Year Plan laid a key emphasis on promoting sustainable development with environmental protection and further improvements in living standards.[7]

In order to assess the efficiency and effectiveness of a development policy with respect to environmental sustainability, environmental factors should be included in the national income system to make it account also for changes in stocks of natural sources and the environment. In this chapter, we have followed the idea of Arrow et al. (2012) in forming the theoretical framework, and after that, presented some preliminary empirical results related to China's (potential) environmentally sustainable growth. The focus of Arrow et al. (2012)'s idea was a broadened concept of wealth. The idea has taken account of both levels and changes in natural capital, human capital, reproducible capital, health capital and knowledge capital in its theoretical framework. We have held our empirical discussion in the light of the very central idea of comprehensive wealth. Section 3.2 introduced the concept of comprehensive wealth and briefly discussed sustainable development from the perspective of comprehensive wealth. After that in Sect. 3.3 we presented and discussed some empirical results in relation to China's potential sustainable growth.

---

[7] For more discussions, see, for example, Zhang and Crooks (2011).

References

Arrow, Kenneth J., Partha Dasgupta, Lawrence H. Goulder, Kevin J. Mumford, and Kirsten Oleson. 2012. Sustainability and the measurement of wealth. *Environment and Development Economics* 17(3): 317–353.

Becker, R. 1982. Intergenerational equity: The capital environment trade-off. *Journal of Environmental Economics and Management* 9: 165–185.

Bovenberg, A.L., and S. Smulders. 1995. Environmental quality and pollution augmenting technological change in a two sector endogenous growth model. *Journal of Public Economics* 57: 369–391.

Brandt, Loren, and Xiaodong Zhu. 2010. *Accounting for China's Growth*. Working Papers tecipa-394, Department of Economics, University of Toronto.

Brock, William A. 1973. Polluted golden age. In *Economics of Natural and Environmental Resources*, ed. V.L. Smith. New York: Gordon & Breach.

Brock, W.A. and M. Scott Taylor. 2005. Economic growth and the environment: A review of theory and empirics. In *The Handbook of Economic Growth*, eds. Aghion P. and Durlauf S, a volume in the series Handbooks in Economics, eds. Kenneth Arrow and Michael D. Intrilligator, Elsevier Publisher.

Brock, William A. and M. Scott Taylor. 2004. *The Green Solow Model*. NBER Working Paper No. 10557.

Chow, Gregory C. 2008. Another look at the rate of increase in TFP in China. *Journal of Chinese Economic and Business Studies* 6(2): 219–224.

Chow, Gregory C., and Kui-Wai Li. 2002. China's economic growth: 1952–2010. *Economic Development and Cultural Change* 51(1): 247–256.

Congressional Budget Office. 2001 CBO's method for estimating potential output: An update. August 2001.

Considine, T.J., and D.F. Larson. 2006. The environment as a factor of production. *Journal of Environmental Economics and Management* 52(3): 645–662.

Copeland, B.R., and M.S. Taylor. 1994. North-south trade and the global environment. *Quarterly Journal of Economics* 109: 755–787.

Easterly, W., and R. Levine. 2001. What have we learned from a decade of empirical research on growth? It's not factor accumulation: Stylized facts and growth models. *The World Bank Economic Review* 15(2): 177–219.

Food and Agriculture Organization. 2006. Global forest resources assessment 2005. United Nations FOA Forestry Paper, 147.

Hall, R.E., and C.I. Jones. 1999. Why do some countries produce so much more output per worker than others? *Quarterly Journal of Economics* 114: 83–116.

Hamilton, Kirk. 2012. Comments on Arrow et al., 'sustainability and the measurement of wealth'. *Environment and Development Economics* 17(3): 356–361.

Houghton, R.A. 2005. Aboveground forest biomass and the global carbon balance. *Global Change Biology* 11: 945–958.

Jiang, Yanqing. 2011. Understanding openness and productivity growth in China: An empirical study of the Chinese provinces. *China Economic Review* 22(3): 290–298.

Jiang, Yanqing. 2012. Technology diffusion, spatial effects and productivity growth in the Chinese provinces. *International Review of Applied Economics* 26(5): 643–656.

Jiang, Yanqing. 2015. Total factor productivity, pollution and 'green' economic growth in China. *Journal of International Development* 27(4): 504–515.

Klenow, Peter, and Andres Rodriguez-Clare. 1997. The neoclassical revival in growth economics: Has it gone too far? *NBER Macroeconomics Annual* 12: 73–103.

Mohtadi, H. 1996. Environment, growth and optimal policy design. *Journal of Public Economics* 63: 119–140.

Musso, Alberto, and Thomas Westermann. 2005. Assessing potential output growth in the euro area: A growth accounting perspective. *ECB Occasional Paper* No. 22, European Central Bank, January 2005.

Nordhaus, W.D., and J. Boyer. 2000. *Warming the World: Economic Models of Global Warming*. Cambridge, MA: MIT Press.

Smulders, S., and R. Gradus. 1996. Pollution abatement and long term growth. *European Journal of Political Economy* 12: 505–532.

Solow, Robert. 2012. A few comments on 'sustainability and the measurement of wealth'. *Environment and Development Economics* 17(3): 354–355.

Tahvonen, O., and J. Kuuluvainen. 1993. Economic growth, pollution and renewable resources. *Journal of Environmental Economics and Management* 24: 101–118.

Tol, R.S.J. 2009. The economic effects of climate change. *Journal of Economic Perspectives* 23(2): 29–51.

Tzouvelekas, E., D. Vouvaki, and A. Xepapadeas. 2006. *Total Factor Productivity Growth and the Environment: A Case For Green Growth Accounting*. Working Papers 0617, University of Crete, Department of Economics.

World Bank. 2006. *Where is the Wealth of Nations? Measuring Capital for the 21st Century*. Washington, DC: World Bank.

World Bank. 2011. *The Changing Wealth of Nations: Measuring Sustainable Development in the New Millennium*. Washington, DC: World Bank.

Xepapadeas, A. 2005. Economic growth and the environment. In *The Handbook of Environmental Economics, Economy wide and International Environmental Issues*, eds. Mäler, K.G. and J. Vincent, a volume in the series Handbooks in Economics, eds. Kenneth Arrow and Michael D. Intrilligator, Elsevier Publishers.

Zhang, Jun. 2008. Estimation of China's provincial capital stock (1952–2004) with applications. *Journal of Chinese Economic and Business Studies* 6(2): 177–196.

Zhang, Jun, Guiying Wu, and Jipeng Zhang. 2007. *Estimating China's Provincial Capital Stock*. Working Paper Series, China Center for Economic Studies, Fudan University.

Zhang, Qingfeng and Robert Crooks. 2011. Environmental Strategy for the 12th Five-Year Plan Period: What Can the People's Republic of China Learn from the 11th Five-Year Plan? *ADB Briefs* No 8, June 2011.

Zheng, Jinghai, Angang Hu, and Arne Bigsten. 2009. Measuring potential output in a rapidly developing economy: The case of China in comparison with the US and EU. Federal Reserve Bank of St. Louis Review, 317–342, July/August 2009.

Chapter 4
Development and the Environmental Kuznets Curve in China

4.1 Introduction

As already mentioned in previous chapters, there prevail considerable concerns about the bad environmental consequences of rapid economic development in China. It has been widely argued that China's economic development has been relying too much on environmental inputs, and the excessive use of environmental inputs is posing serious threats to China's water, air, bio-system and energy supplies (Co et al. 2008), leading to bad consequences such as environmental deterioration and resource degradation. Total direct losses from such environmental damages were estimated to account for ten percent of total income of China annually (the Economist 2004; Cheng et al. 2007.

The issue of the relationship between economic performance and the quality of the environment in China has been tackled, to varying degrees, by some existing literature, a part of which followed the framework of the Environmental Kuznets Curve (EKC). Early EKC literature usually aimed at exploring the effect of economic performance on environmental quality (see, for example, Selden and Song 1994; Vincent 1997; Gale and Mendez 1998).[1] The EKC exhibits an inverted U-shaped relation between environmental pollution and per capita income. That is to say, increased income is associated with an increase in pollution in less developed economies but a decrease in pollution in more developed economies (see, for example, Grossman and Krueger 1993, 1995). Unavoidably, A simple investigation of the effect of economic performance on environmental quality tends to overlook the fact that the linkage between income growth or level (on the one hand) and

[1]For a broader range of literature on economic performance and the environment, see, for example, Brock (1973), Becker (1982), Tahvonen and Kuuluvainen (1993), Copeland and Taylor (1994), Bovenberg and Smulders (1995), Mohtadi (1996), Smulders and Gradus (1996), Vincent (1997), Nordhaus and Boyer (2000), Brock and Taylor (2004, 2005), Xepapadeas (2003, 2005), Tzouvelekas and Xepapadeas (2006), Considine and Larson (2006), Tol (2009), World Bank (2006, 2011), Zhang and Crook (2011), Arrow et al. (2012), and Solow (2012).

Y. Jiang, *Green Development in China*,
SpringerBriefs in Economics, DOI 10.1007/978-981-10-0693-7_4

environmental pollution (on the other) may actually vary depending on the source of income growth, as growth in different industrial sectors may bear different pollution intensities (Abdulai and Ramcke 2009; Antweiler et al. 2001).

In this chapter, we provide a preliminary discussion of economic development and the EKC in the case of China. This chapter serves as a precursor study to later chapters. In subsequent chapters we will come back to the issue of the relationship between economic performance (as well as other factors related to economic performance) and environmental quality. The structure of this chapter goes as follows. In Sect. 4.2, we present the theoretical foundation, which is supposed to shed light on our subsequent discussions in later sections of this chapter as well as in later chapters. In Sect. 4.3 we provide some discussion on descriptive statistics in order to gain an intuitive and heuristic idea about the issue under study in this chapter. Section 4.4 is devoted to further discussion of the EKC issue, where various estimation results based on a series of regression exercise are referred to from time to time. Finally, we conclude this chapter in Sect. 4.5.

4.2 The Theoretical Foundation

Various types of optimization models, static or dynamic, usually serve as the underlying theoretical foundation for the income-pollution relationship that constitutes the focus of the EKC studies. To incorporating environmental factors into a theoretical growth framework, we can simply follow Xepapadeas (2003) closely and consider a theoretical setup where a flow of pollution and an effort of abatement feature in. For this purpose, we assume that an emission function of the form $Z = v(k, a)$ describes the flow of pollution generated along time, where a represents abatement at time t. The optimization problem can be captured by

$$\begin{aligned} &\max_{\{c(t)\}} \int_0^\infty e^{-\rho t}[U(c) - D(Z)]dt,\ D' > 0,\ D'' < 0 \\ &\text{s.t.}\ \dot{k} = f(k) - c - a - \delta k \end{aligned} \tag{4.1}$$

where the function $f(k)$ is the standard neoclassical production function and obviously we have assumed zero population growth and technology change. The current value Hamiltonian is then written as

$$H = U(c) - D(v(k, a)) + q(f(k) - c - a - \delta k) \tag{4.2}$$

where the first-order conditions are

$$U_c(c) = q,\ D_{v_a}(k, a) = q \tag{4.3}$$

from which we can derive the following dynamic system describing the evolution of the economy in terms of (c, k)

$$\dot{c}/c = \frac{1}{\eta}\left[f'(k) - \frac{D'_{v_k}(k, \tilde{a}(c,k))}{U_c(c)} - \rho - \delta\right] \tag{4.4}$$

$$\dot{k} = f(k) - c - \tilde{a}(c,k) - \delta k \tag{4.5}$$

where $a = \tilde{a}(c,k)$ is the short-run optimal level of abatement derived from (4.3). Assuming that a steady state (c^*, k^*) exists with the saddle point having one stable arm converging to it, we can obtain the policy function $c = \hat{c}(k)$ as traced out by the stable arm. Given this, the pollution-income relationship (*PIR*) can be defined by

$$PIR = \{(Z,y) : Z = v(k, \tilde{a}(\hat{c}(k),k)), y = f(k)\} \tag{4.6}$$

Using (4.4) and (4.5), we end up with the following

$$\hat{c}'(k) = \frac{\frac{1}{\eta}\left[f'(k) - \frac{D'_{v_k}(k,\tilde{a}(\hat{c}(k),k))}{U_c(\hat{c}(k))} - \rho - \delta\right]\hat{c}(k)}{f(k) - \hat{c}(k) - \tilde{a}(\hat{c}(k),k) - \delta k} \tag{4.7}$$

for which $\hat{c}'(k^*)$ specifically represents the slope of the stable arm at the steady state.

It should be noted that (4.6) above describes the pollution-income relationship on the socially optimal path of the economy. However, economies for which empirical data are used to estimate the pollution-income relationship may not necessarily be on any socially optimal path. The shape of the pollution-income relationship in fact depends on preference, technology and regulation parameters. In this sense, the shape of the empirically estimated pollution-income relationship could reflect, based on the sample period, whether the economy is following an optimal path, or whether environmental policy intervenes to change the shape of the pollution-income relationship.

Alternatively, similar results can be derived from a simpler model of a static one producer-consumer style by Levinson (2002) (see also Xepapadeas 2003), where consumption c is the source of pollution. Then utility is written as $U = c - Z$ and pollution is described by $Z = c - c^{\xi}a^{\varphi}$, where a again denotes the abatement effort and $c^{\xi}a^{\varphi}$ is pollution abatement. The optimization problem can thus be written as the following

$$\max_{(c,a)} c - (c - c^{\xi}a^{\varphi})$$
$$\text{s.t. } c + a = y \tag{4.8}$$

The pollution-income relationship is then described by

$$Z(y) = c(y) - [c(y)]^{\xi}[a(y)]^{\varphi} \tag{4.9}$$

where $c(y) = \xi/(\xi + \varphi)$, $a(y) = \varphi/(\xi + \varphi)$. Whether the pollution-income relationship follows the inverted U shape or is simply a straight line (with a positive slope) depends on whether $\xi + \varphi$ is greater than or equal to unity. This modeling can be easily extended to a multi-person situation, where any representative individual solve the following problem

$$\begin{aligned} &\max_{(c_i, a_i)} c_i - (c - c^{\xi}a^{\varphi}) \\ &\text{s.t. } c_i + a_i = y_i,\ c = \sum_{i=1}^{n} c_i,\ a = \sum_{i=1}^{n} a_i \end{aligned} \tag{4.10}$$

where, again, the pollution-income relationship has an inverted U shape if $\xi + \varphi > 1$. The same result (of the inverted U shape of the pollution-income relationship) holds for the planner's problem when $\xi + \varphi > 1$

$$\max_{(c_i, a_i)} \sum_{i=1}^{n} c_i - nZ \tag{4.11}$$

Therefore, the technology parametrization $\xi + \varphi > 1$ indicates that the inverted U-shaped pollution-income relationship is consistent with both efficient resource allocation in the case of the social planner's problem and market failure in the case of multiple agents.

4.3 Descriptive Statistics

The key point expressed by the theoretical modeling above is that the economy is confronted with a tradeoff between goods production and environmental quality, given a certain level of technology. Before we formally discuss the empirical modeling and carry out our econometric analysis, it is helpful that we first focus on some descriptive statistics concerning pollution and income across the difference Chinese regions over time.

First, to get a feel for the interrelations of representative regional pollution variables and income-related variables, we first display some basic descriptive statistics regarding three representative pollution variables and two income-related variables across different regions in China using data for the year 2013. These are contained in Table 4.1. The first three variables listed in Table 4.1 are main regional pollutant emissions in waste gas (in ten thousand tons). These pollutants are namely sulfur dioxide (sul), nitrogen oxides (nit), and smoke and dust (smo). The last two variables in Table 4.1 are regional income-related variables, namely, gross regional product (grp) and gross fixed capital formation (gfc), both are nominal values

Table 4.1 Descriptive statistics of five representative variables

Variable	Obs	Mean	Std. Dev.	Min	Max
sul	31	65.93	41.36	0.42	164.50
nit	31	71.85	45.23	4.43	165.25
smo	31	41.23	30.12	0.68	131.33
grp	31	20322.88	15597.43	807.67	62163.97
gfc	31	11358.46	7336.98	898.40	29249.50

Table 4.2 Correlations between the Variables

Variable	sul	nit	smo	grp	gfc
sul	1				
nit	0.9055	1			
smo	0.8139	0.8354	1		
grp	0.4968	0.6680	0.2727	1	
gfc	0.6715	0.8099	0.4442	0.9366	1

measured in 100 million yuan. Given our theoretical discussion in the preceding section, we now expect that the representative variables listed in Table 4.1 follow a certain pattern of interrelationship.

To see this, we conduct a correlation test among the five variables listed in Table 4.1. The results are contained in Table 4.2, which are measured in correlation coefficients. We can easily see that all the coefficients have the expected positive sign, which is not surprising. Seeing that the three major pollutants are highly correlated, we can construct an 'overall' pollution emission measure simply by taking the sum of the emitted quantities of the three pollutants. We suspect that there is a relationship between the regional level of income (as measured by gross regional product per capita) and overall regional pollution emission per capita (as measured here by the total emitted quantity of the three pollutants). Therefore, we display the relationship (across the 31 Chinese regions for the specific year of 2013) in a scatterplot in Fig. 4.1, where the horizontal axis depicts gross regional product measured in 100 million yuan and the vertical axis depicts the total emitted quantity of the three representative pollutants measured in ten thousand tons. As the plot does not control for any region heterogeneity, we cannot see any discernable pattern of an inverted U-shaped relationship between the two variables.

Next, we depict the time trend of regional pollution emission using time series data on a typical pollutant (sulfur dioxide, measured in ten thousand tons). The trends for seven representative regions (provinces) are chosen and sketched in Fig. 4.2. As most Chinese regions follow a clear ascending trend for regional income and regional per capita income over time, we expect that, if the EKC applies, then regional pollution emission should follow an inverted-U shaped trend over time. We see in Fig. 4.2 that for all of the seven regions included, regional pollution (sulfur dioxide) emission follows a clear time trend that has a typical inverted-U shape during 2003–2010.

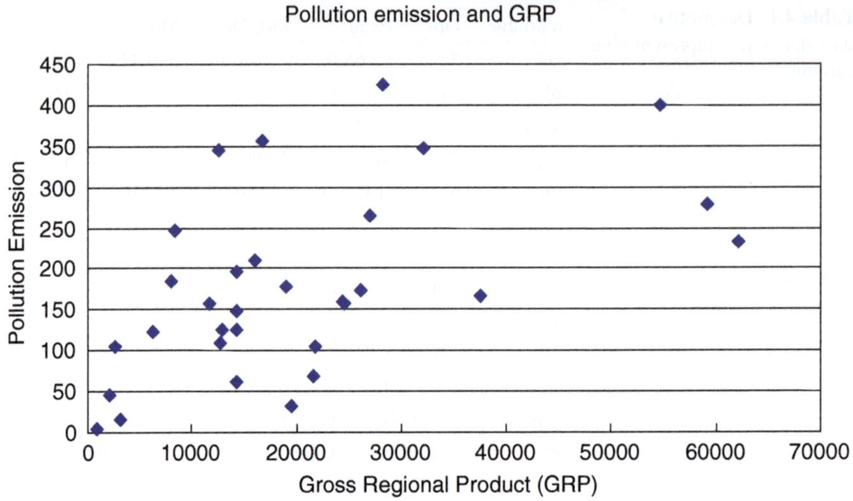

Fig. 4.1 Pollution emission and gross regional product

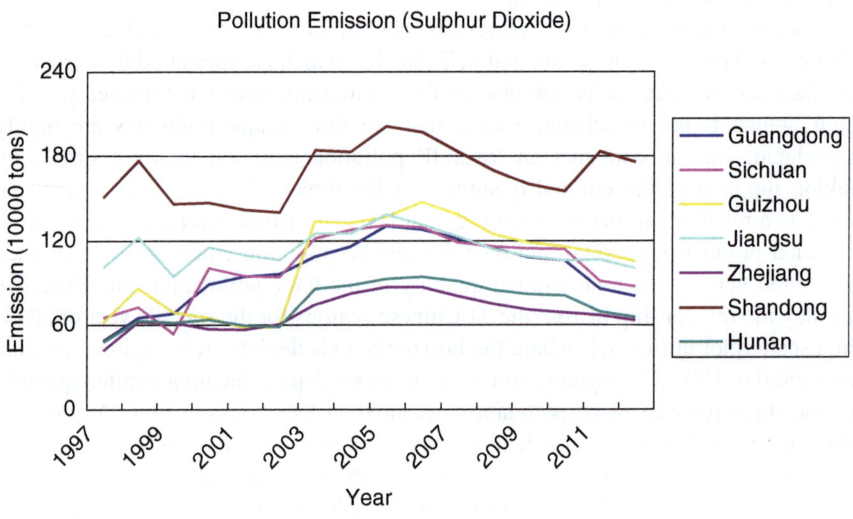

Fig. 4.2 Regional pollution emission by year

4.4 Discussion of Some Empirical Results

In several of our previous studies, such as Jiang (2014, 2015), we approached the issue of the environmental consequences of economic development in China from different angles. For example, as already mentioned in the previous chapter, Jiang (2015) aimed to explore the relationship between TFP growth and pollution emissions across the Chinese provinces. Specifically, the study aimed to investigate whether the use of the environment as a production input substantially contributed to output growth in the case of the Chinese regions. Some other previous studies of ours, however, touched upon the EKC issue. One major point is that these studies realized that an economy faces a tradeoff between goods production and environmental quality, given a certain level of technology. Therefore, it is natural to think that for environmentally sustainable development to take place, technological advancement in goods production is required in order to facilitate increases in per capita income, and on the other hand, technological advancement in tackling pollution emission must exceed pollution accumulation in order for pollution to abate and the environment to improve (see also Brock and Taylor 2004, 2005).

In one of our previous studies, our empirical results showed that with regional population controlled for in our regressions, as regional income or regional per capita income grows larger, pollution emission and emission intensity first increases and then declines. This result supported the inverted U-shaped EKC hypothesis. Our empirical results also showed that once regional per capita income, regional population, and regional per capita physical capital and human capital stocks, as well as the individual region heterogeneity were controlled for, regional openness to foreign trade exerted a positive partial effect on regional per capita pollution emission. It was likely that regional openness to foreign trade may affect regional pollution emission by affecting regional input of natural resources and regional industry mix, where a change in either of the latter two may be associated with a change in regional TFP.

According to our empirical results, once the effects of a host of other explanatory variables were partialed out, regional physical capital accumulation was not shown to have a significant partial effect on regional pollution emission. However, it was shown that ceteris paribus, a higher level of regional per capita human capital corresponded to a lower level of regional per capita pollution emission. Also, when the partial effects of some other explanatory variables were netted out, a change in regional population was associated with a change in either regional TFP or/and the level of regional resource input, both of which may in turn exert effects on regional pollution emission. Our empirical results supported this point. Our results also supported the EKC hypothesis by showing that as TFP hence (per capita) income grows larger, pollution emission and emission intensity first increases and then declines.

Likewise, in another previous study of ours, empirical results showed that controlling for regional output, regional population, and the levels of regional domestically invested physical capital stock and human capital stock, along with the individual region heterogeneity, regional FDI had a positive partial effect on regional pollution emission. The possible mechanism may be that when regional output, population, and the levels of domestically invested physical capital stock and human capital stock are controlled for, regional FDI may affect regional pollution emission through affecting either the regional input of natural resources or the regional industry mix, where a change in either may be associated with a change in the level of regional TFP.

4.5 Concluding Remarks

The relationship between economic performance and the quality of the environment has been of interest in literature. Some of such literature explored the effect of economic performance on environmental quality within the framework of the EKC. The EKC traces out an inverted U-shaped relationship between environmental pollution and per capita income: increased income is accompanied by an increase in pollution in less developed economies but a decrease in pollution in more developed economies. However, the mere use of the EKC to describe the effect of economic performance on environmental quality is kind of simplistic. A simple investigation of the relationship between economic performance and environmental quality is likely to neglect the fact that the linkage between income growth or level (on the one hand) and environmental pollution (on the other) may actually differ according to the source of income growth. In this chapter, we have provided a precursor discussion of economic development and the EKC in the case of China. In this sense, this chapter opens a prelude to the subsequent chapters, where we will begin to study the related issues in greater depth.

Among other findings, our empirical results showed that with regional population controlled for in our regressions, as regional income or regional per capita income grows larger, pollution emission and emission intensity first increases and then declines. This result lent support to the inverted U-shaped EKC hypothesis. Moreover, our empirical results also showed that physical capital accumulation did not have a significant effect on pollution emission, but a higher level of human capital intensity helped contribute to a lower level of per capita pollution emission.

References

Abdulai, Awudu and Linda Ramcke. 2009. The impact of trade and economic growth on the environment: revisiting the cross-country evidence. Kiel Working Paper No. 1491, Kiel Institute for the World Economy, March 2009.

Antweiler, W., B.R. Copeland, and M.S. Taylor. 2001. Is free trade good for the environment? *American Economic Review* 91(4): 877–908.

Arrow, Kenneth J., Partha Dasgupta, Lawrence H. Goulder, Kevin J. Mumford, and Kirsten Oleson. 2012. Sustainability and the measurement of wealth. *Environment and Development Economics* 17(3): 317–353.

Becker, R. 1982. Intergenerational equity: The capital environment trade-off. *Journal of Environmental Economics and Management* 9: 165–185.

Bovenberg, A.L., and S. Smulders. 1995. Environmental quality and pollution augmenting technological change in a two sector endogenous growth model. *Journal of Public Economics* 57: 369–391.

Brock, William A. 1973. Polluted Golden Age. In *Economics of natural and environmental resources*, ed. V.L. Smith. New York: Gordon & Breach.

Brock, William A. and M. Scott Taylor. 2004. The green solow model. NBER Working Paper No. 10557.

Brock, W.A. and M. Scott Taylor. 2005. *Economic Growth and the Environment: A Review of Theory and Empirics*. *The Handbook of Economic Growth* (P. Aghion and S. Durlauf, Eds), a volume in the series Handbooks in Economics, edited by, Kenneth Arrow and Michael D. Intrilligator, Elsevier Publisher.

Cheng, Y., S. Chen and Y. Guang. 2007. Expected decisions from the two meetings. *Liaowang Dongfang Weekly Journal*.

Co, C.Y., F. Kong and S. Lin. 2008. Pollution across chinese provinces. Development Economics Working Papers 22148, East Asian Bureau of Economic Research.

Considine, T.J., and D.F. Larson. 2006. The environment as a factor of production. *Journal of Environmental Economics and Management* 52(3): 645–662.

Copeland, B.R., and M.S. Taylor. 1994. North-south trade and the global environment. *Quarterly Journal of Economics* 109: 755–787.

Economist, The. 2004. China's environment: A great wall of waste.

Gale, L.R., and J.A. Mendez. 1998. The empirical relationship between trade, growth and the environment. *International Review of Economics and Finance* 7(1): 53–61.

Grossman, G.M. and A.B. Krueger. 1993. Environmental impacts of the North American free trade agreement. *The U.S.-Mexico Free Trade Agreement*, P. Garber. Massachusetts Institute of Technology, The MIT Press, Cambridge, MA., 13–56.

Grossman, G.M., and A.B. Krueger. 1995. Economic growth and the environment. *Quarterly Journal of Economics* 110(2): 353–377.

Jiang, Yanqing. 2014. Openness, economic development, and the environment in post-reform China: a model with empirical analysis. *International Journal of Social Science Studies* 2(3): 56–66.

Jiang, Yanqing. 2015. Total factor productivity, pollution and 'green' economic growth in China. *Journal of International Development* 27(4): 504–515.

Levinson, A. 2002. The ups and downs of the environmental Kuznets curve. In *Recent Advances in Environmental Economics*, ed. J. List, and A. de Zeeuw. Cheltenham: Edgar Elgar.

Mohtadi, H. 1996. Environment, growth and optimal policy design. *Journal of Public Economics* 63: 119–140.

Nordhaus, W.D., and J. Boyer. 2000. *Warming the world: Economic models of global warming*. Cambridge: MIT Press.

Selden, T.M., and D. Song. 1994. Environmental quality and development: Is there a kuznets curve for air pollution emissions? *Journal of Environmental Economics and management* 27: 147–162.

Smulders, S., and R. Gradus. 1996. Pollution abatement and long term growth. *European Journal of Political Economy* 12: 505–532.

Solow, Robert. 2012. A few comments on 'sustainability and the measurement of wealth'. *Environment and Development Economics* 17(3): 354–355.

Tahvonen, O., and J. Kuuluvainen. 1993. Economic growth, pollution and renewable resources. *Journal of Environmental Economics and Management* 24: 101–118.

Tol, R.S.J. 2009. The economic effects of climate change. *Journal of Economic Perspectives* 23(2): 29–51.

Tzouvelekas, E., D. Vouvaki, and A. Xepapadeas. 2006. Total factor productivity growth and the environment: A case for green growth accounting. Working Papers 0617, University of Crete, Department of Economics.

Vincent, J.R. 1997. Testing for environmental kuznets curves within a developing country. *Environment and Development Economics* 2: 417–431.

World Bank. 2006. *Where is the wealth of nations? Measuring capital for the 21st century*. Washington: World Bank.

World Bank. 2011. *The changing wealth of nations: Measuring sustainable development in the New Millennium*. Washington: World Bank.

Xepapadeas, A. 2003. Economic growth and the environment. Karl Göran Mäler and Jeffrey Vincent (eds), *Handbook of Environmental Economics*.

Xepapadeas, A. 2005. Economic growth and the environment. *Handbook of Environmental Economics*, Economy wide and International Environmental Issues, (Mäler, K.G. and J. Vincent Eds), a volume in the series Handbooks in Economics, edited by, Kenneth Arrow and Michael D. Intrilligator, Elsevier Publishers.

Zhang, Qingfeng and Robert Crooks. 2011. Environmental strategy for the 12th five-year plan period: What Can the People's Republic of China Learn from the 11th Five-Year Plan? *ADB Briefs* No 8, June 2011.

Chapter 5
Economic Growth and Environmental Input

5.1 Introduction

More than a decade ago in 2003, the Chinese Communist Party proposed to follow a human-oriented approach and promote social advancement by achieving comprehensive, coordinated, and sustainable economic development. Five balanced plans were proposed: balancing development between urban and rural areas, between different geographical regions, between economic and social undertakings, between humanity and nature, and between domestic development and opening up to the outside world. The Chinese government established the largest green development program in its 12th Five-Year Plan (2011–2015).

The government's efforts to engage in green development reflected the fact that there had been bad environmental consequences of rapid economic development in China. As already mentioned in earlier chapters, China's economic development was criticized to be relying too much on environmental factors, and the excessive use of environmental inputs was posing serious threats to China's environmental resources, resulting in bad consequences such as environmental deterioration and resource degradation.

Existing literature has dealt with different aspects of the various environmental issues of economic development to various degrees. Some literature has traced an Environmental Kuznets Curve (EKC) (see again, for example, Selden and Song 1994; Vincent 1997; Gale and Mendez 1998).[1] As discussed in earlier chapters, the

[1]The same as for the previous two chapters, the reader can refer to, for example, Brock (1973), Becker (1982), Tahvonen and Kuuluvainen (1993), Grossman and Kruger (1993, 1995), Copeland and Taylor (1994), Bovenberg and Smulders (1995), Mohtadi (1996), Smulders and Gradus (1996), Vincent (1997), Nordhaus and Boyer (2000), Antweiler et al. (2001), Levinson (2002), Brock and Taylor (2004, 2005), Xepapadeas (2003, 2005), Tzouvelekas et al. (2006), Considine and Larson (2006), Co et al. (2008), Abdulai and Ramcke (2009), Tol (2009), World Bank (2006, 2011), Zhang and Crook (2011), Arrow et al. (2012), and Solow (2012), Jiang (2014, 2015) for a broader review of the related literature for the issue in question.

Y. Jiang, *Green Development in China*,
SpringerBriefs in Economics, DOI 10.1007/978-981-10-0693-7_5

EKC exhibits an inverted U-shaped relation between environmental pollution and per capita income. In the prior chapter, we presented a pilot discussion of economic development and the EKC in the case of the Chinese regions. Our empirical results in the preceding chapter showed that with some other variables properly controlled for, as regional income or regional per capita income grows larger, pollution emission and emission intensity first increases and then declines. The China case study thus lent support to the inverted U-shaped EKC hypothesis proposed by earlier literature.

In this chapter, we therefore approach the issue from a somewhat different angle. We will present an empirical discussion of the relationship between economic growth and the environmental input, based on related empirical evidence from the Chinese regions. The study in this chapter explores whether the use of environmental resources taken as a factor of production would contribute crucially to regional economic growth in the case of the Chinese regions. Empirical results in this chapter show that for the Chinese regions a significant part of output growth and productivity growth could actually be possibly attributed to the increasing use of environmental inputs. The organization of this chapter is as follows. In Sect. 5.2, we present the theoretical setup, which informs and guides subsequent empirical discussions in later sections of this chapter. In Sect. 5.3 we move on to the econometric modeling, which serves as the technical framework for our empirical analysis coming next. In Sect. 5.4 we describe the sample and data, and present and discuss our regression results. Finally, we conclude this chapter in Sect. 5.5.

5.2 The Theoretical Modeling

Before we are able to explore the issue empirically, we need to build a theoretical framework as the foundation for our empirical analysis. In one of our previous studies, Jiang (2015), by following Xepapadeas (2003), we considered an extended version of the Solow growth model that takes the flow of pollution emissions as a factor of production. We thus reproduce the setup of the theoretical model here. In this model, the production function for the aggregate economy takes the form

$$Y = K^{\alpha}(AL)^{\beta}(BZ)^{\gamma} \tag{5.1}$$

in which Y is the level of output, K is the level of capital stock, L is the level of labor input, A is the level of labor-augmenting technology, Z stands for the level of a flow of pollution emissions that can be kept constant at different levels, which are dependent on policy considerations and technological conditions. The production function in (5.1) thus conveniently takes pollution emissions as a factor of production, where the coefficient B is the level of technology that augments the efficiency of pollution emissions as a production input. Constant returns to scale is assumed with respect to K, AL and BZ, that is to say, the structural parameters α, β and γ sum to unity. To keep the model simple and tractable, we further assume that L, A and B grow exogenously at rates n, g and b, respectively, that is,

$g_L \equiv \dot{L}/L = n$, $g_A \equiv \dot{A}/A = g$ and $g_B \equiv \dot{B}/B = b$, where a dot overhead denotes the time derivative.[2]

By making use of the constant-returns-to-scale structure of (5.1), we can then rewrite the production function intensively in the per worker (per unit of labor) form as

$$y = k^{\alpha} A^{\beta} (Bz)^{\gamma} \tag{5.2}$$

where we define $y \equiv Y/L$, $k \equiv K/L$ and $z \equiv Z/L$. For simplicity, we can assume that each worker owns one unit of labor. By taking logs it is easy to see that Eq. (5.2) in turn implies that the growth rates follow this relationship

$$g_y = \alpha g_k + \beta g + \gamma(b - n) \tag{5.3}$$

where, for simplicity, we temporarily fix the level of the flow of pollution emissions Z at a constant level. We can now show that the economy converges to a steady state (or balanced growth path). By defining the capital-output ratio as $\psi \equiv K/Y = k/y$, Eq. (5.2) implies that the capital-output ratio can be written as

$$\psi = k^{1-\alpha} A^{-\beta} (Bz)^{-\gamma} \tag{5.4}$$

which further yields

$$g_{\psi} = (1 - \alpha) g_k - \beta g - \gamma(b - n) \tag{5.5}$$

in which the relation $g_k = g_K - n$ necessarily holds by construction.

The dynamic evolution of K follows $\dot{K} = sY - \delta K$, in which s and δ are the investment rate and depreciation rate respectively. Therefore we end up with this relation

$$g_k = g_K - n = \frac{s}{\psi} - \delta - n \tag{5.6}$$

Inserting (5.6) back into (5.5) yields

$$\dot{\psi} = (1 - \alpha)s - \lambda\psi \tag{5.7}$$

in which λ is defined as $\lambda \equiv (1 - \alpha)\delta + \beta(n + g) + \gamma b > 0$. It then can be easily seen that ψ converges to a steady state value of

$$\psi^* = \frac{(1 - \alpha)s}{\lambda} \tag{5.8}$$

---

[2]That is, we conveniently use g_X ($\equiv \dot{X}/X$) to denote the growth rate of variable X over time for any variable X.

which simply shows that on the balanced growth path, per worker capital k and per worker output y grow at the same rate. Referring back to Eq. (5.3), we end up having the steady state growth rate of per worker capital and per worker output

$$g_k^* = g_y^* = \frac{\beta g + \gamma(b-n)}{1-\alpha} \equiv \xi \tag{5.9}$$

in which, for convenience sake, the value of the growth rate is denoted ξ. Similarly, the growth rates of K and Y (where $g_K = g_k + n$ and $g_Y = g_y + n$ by construction) also take the same following value on the balanced growth path, that is

$$g_K^* = g_Y^* = \frac{\beta(n+g) + \gamma b}{1-\alpha} = \xi + n \tag{5.10}$$

With a bit of rearrangement, the production function in (5.1) becomes

$$Y = \psi^{\alpha/(1-\alpha)}(AL)^{\beta/(1-\alpha)}(BZ)^{\gamma/(1-\alpha)} \tag{5.11}$$

We further define

$$\tilde{y} \equiv Y \Big/ \left[(AL)^{\beta/(1-\alpha)}(BZ)^{\gamma/(1-\alpha)}\right] \tag{5.12}$$

$$\tilde{k} \equiv K \Big/ \left[(AL)^{\beta/(1-\alpha)}(BZ)^{\gamma/(1-\alpha)}\right] \tag{5.13}$$

By referring back to the production function (5.1), we have

$$\tilde{y} = \tilde{k}^{\alpha} \tag{5.14}$$

Therefore the dynamic evolution of K (i.e. $\dot{K} = sY - \delta K$) implies that the steady state in terms of $\tilde{k}$ and $\tilde{y}$ can be expressed as

$$\tilde{k}^* = \left[\frac{(1-\alpha)s}{\lambda}\right]^{1/(1-\alpha)} \tag{5.15}$$

$$\tilde{y}^* = \left[\frac{(1-\alpha)s}{\lambda}\right]^{\alpha/(1-\alpha)} \tag{5.16}$$

where it should be noted that $\lambda/(1-\alpha) = \delta + \xi + n$.

Therefore, for any specific level of Z, it is easily shown that on the balanced growth path, aggregate output grows by following this time path

$$Y^*(t;Z) = \left[\frac{(1-\alpha)s}{\lambda}\right]^{\alpha/(1-\alpha)} (A_0L_0)^{\beta/(1-\alpha)}(B_0Z)^{\gamma/(1-\alpha)}e^{(\xi+n)t} \tag{5.17}$$

in which the initial values of A, L and B are denoted A_0, L_0 and B_0. By the same token, on the balanced growth path, per worker output grows according to

$$y^*(t;Z) = \left[\frac{(1-\alpha)s}{\lambda}\right]^{\alpha/(1-\alpha)} A_0^{\beta/(1-\alpha)}(B_0Z/L_0)^{\gamma/(1-\alpha)}e^{\xi t} \tag{5.18}$$

where we can see that $\partial y^*/\partial Z > 0$ and $\partial^2 y^*/\partial Z^2 < 0$. Thus, other things equal, an increase in the pollution emissions input raises the steady state level of (per worker) output at a decreasing rate. Since the variable Z can be thought of as a policy variable, the level of Z can thus be chosen according to considerations on the tradeoff between economic growth and the environmental quality.

We further assume that the stock of pollution, denoted D, accumulates according to

$$\dot{D} = Z - \rho D \tag{5.19}$$

where $\rho > 0$ is the natural rate of regeneration (and that D equals zero represents a pristine environment with no pollution stock). For the environmental quality to improve, the level of the policy variable Z should be chosen (at each point in time) so that $\dot{D} < 0$. If we assume that on the balanced growth path discussed above, the stock of environmental pollution should decrease, say, at a constant rate, then the level of Z should be chosen so that Z decreases at the same rate as D.[3] Suppose that on the balanced growth path, $g_D^* = g_Z^* = -m$, and the level of Z is chosen so that $Z(t) = Z_0e^{-mt}$, where $0 < m < \rho$, then Eq. (5.18) becomes

$$y^*(t) = \left[\frac{(1-\alpha)s}{\lambda}\right]^{\alpha/(1-\alpha)} A_0^{\beta/(1-\alpha)}(B_0Z_0/L_0)^{\gamma/(1-\alpha)}e^{\vartheta t} \tag{5.20}$$

where we define $\vartheta \equiv [\beta g + \gamma(b - n - m)]/(1-\alpha)$, and Z_0 denotes the level of Z at the initial time. Given (5.19) and (5.20), we can see that long-run environmentally sustainable growth requires the following

$$\beta g + \gamma(b - n - m) > 0 \tag{5.21}$$

This inequality implies that in the long run, sufficiently large values of g and b are needed for green economic growth to be sustained.

[3]This is because according to (5.19), $g_D = Z/D - \rho$. Then Z/D must be a constant on the balanced growth path if g_D is assumed to be a constant on the balanced growth path.

5.3 Econometric Setup

Based on the theoretical discussion above, we can now move on to our econometric modeling. We following Jiang (2015), and Linearize around the steady state $\tilde{y}^*$. The speed of convergence of $\tilde{y}$ is then governed by

$$\frac{d\ln\tilde{y}(t)}{dt} = \lambda[\ln\tilde{y}^* - \ln\tilde{y}(t)] \tag{5.22}$$

where $\lambda = (1-\alpha)\delta + \beta(n+g) + \gamma b > 0$, as given above already. Equation (5.22) implies

$$\ln\tilde{y}(t_2) = (1 - e^{-\lambda\tau})\ln\tilde{y}^* + e^{-\lambda\tau}\ln\tilde{y}(t_1) \tag{5.23}$$

where $\tilde{y}(t_1)$ and $\tilde{y}(t_2)$ are values of $\tilde{y}(t)$ at time t_1 and t_2 ($t_1 < t_2$), and $\tau = (t_2 - t_1)$. Further, Eq. (5.23) can expressed in the per worker form as

$$\begin{aligned}\ln y(t_2) = & e^{-\lambda\tau}\ln y(t_1) + \left(1 - e^{-\lambda\tau}\right)\frac{\alpha}{1-\alpha}\ln s - \left(1 - e^{-\lambda\tau}\right)\frac{\alpha}{1-\alpha}\ln(n+\xi+\delta) \\ & + \left(1 - e^{-\lambda\tau}\right)\frac{\gamma}{1-\alpha}\ln Z + \left(1 - e^{-\lambda\tau}\right)\ln V_0 + \xi\left(t_2 - e^{-\lambda\tau}t_1\right)\end{aligned} \tag{5.24}$$

in which $V_0 \equiv A_0^{\beta/(1-\alpha)}(B_0/L_0)^{\gamma/(1-\alpha)}$. A slightly different variant of Eq. (5.24) is written as

$$\begin{aligned}\ln y(t_2) = & e^{-\lambda\tau}\ln y(t_1) + \left(1 - e^{-\lambda\tau}\right)\frac{\alpha}{1-\alpha}\ln s - \left(1 - e^{-\lambda\tau}\right)\frac{\alpha}{1-\alpha}\ln(n+\xi+\delta) \\ & + \left(1 - e^{-\lambda\tau}\right)\frac{\gamma}{1-\alpha}\ln z + \left(1 - e^{-\lambda\tau}\right)\ln W_0 + \chi\left(t_2 - e^{-\lambda\tau}t_1\right)\end{aligned} \tag{5.25}$$

in which we have replaced the aggregate-level variable Z in (5.24) with the corresponding per worker term $z \equiv Z/L$, defining at the same time $W_0 \equiv A_0^{\beta/(1-\alpha)}B_0^{\gamma/(1-\alpha)}$ and $\chi \equiv (\beta g + \gamma b)/(1-\alpha)$. Our empirical analysis in the following section will be based on Eqs. (5.24) and (5.25), where the ultimate choice between the two specifications depends on whether we think the aggregate-level, Z, or the per worker level, z, would be associated with a better policy variable with respect to the emission standard that is assumed to remain relatively stable over the interval $\tau = (t_2 - t_1)$.

In its conventional form, the regression specification of a panel data structure based on (5.25) (or similarly on (5.24)) can be written as

$$\bar{y}_{it} = \pi \bar{y}_{i,t-1} + \sum_{j=1}^{3} \varphi_j x_{it}^j + \eta_t + c_i + \varepsilon_{it} \tag{5.26}$$

in which the subscripts i and t index the cross section units (geographical region) and the time periods respectively. $\bar{y}_{it}$ and $\bar{y}_{i,t-1}$ correspond to $\ln y(t_2)$ and $\ln y(t_1)$, with $\pi \equiv e^{-\lambda\tau}$. The x_{it}^j's represent the three variables $\ln s$, $\ln(n + \xi + \delta)$ and $\ln z$ in (5.25), and the coefficients on them are $\varphi_1 \equiv (1 - e^{-\lambda\tau})\frac{\alpha}{1-\alpha}$, $\varphi_2 \equiv -(1 - e^{-\lambda\tau})\frac{\alpha}{1-\alpha}$, and $\varphi_3 \equiv (1 - e^{-\lambda\tau})\frac{\gamma}{1-\alpha}$. The term c_i is the time-invariant region heterogeneity that accounts for the term $(1 - e^{-\lambda\tau}) \ln W_0$ in (5.25), η_t is the time-variant intercept taking account of the term $\chi(t_2 - e^{-\lambda\tau} t_1)$ in (5.25), and finally ε_{it} is a zero-mean idiosyncratic error that varies across geographical regions and over time periods.

A regression setup like the one above nevertheless generates several practical issues. First, there exists a nuisance variable. The major purpose of our regression analysis is to estimate the unknown values of the structural parameters α and β, but the value of ξ in the explanatory variable $\ln(n + \xi + \delta)$ is unknown in the sense that it depends on the values of α and β whose values are unknown a priori. Second, the value of λ depends on the level of n, which varies across regions and time periods. Seeing these difficulties, we therefore modify our model in (5.26) accordingly to simplify it and make it econometrically tractable. We need to determine the likely value of ξ. In our first round of regression exercise, we temporarily set $\xi = 0.02$. Later we will run more regressions based on different values of ξ as a sensitivity analysis and robustness check. As a result, our baseline regression specification is now written as

$$\begin{aligned}\ln y_{i,t+1} = & e^{-\lambda\tau} \ln y_{it} + \left(1 - e^{-\lambda\tau}\right)\frac{\alpha}{1-\alpha} \ln s_{it} - \left(1 - e^{-\lambda\tau}\right)\frac{\alpha}{1-\alpha} \ln(n_{it} + 0.05) \\ & + \left(1 - e^{-\lambda\tau}\right)\frac{\gamma}{1-\alpha} \ln x_{it} + c_i + \eta_t + \varepsilon_{it}\end{aligned} \tag{5.27}$$

where we have assumed a unified value of the depreciation rate (across regions and time periods) that is $\delta = 0.03$ (see, for example, Mankiw et al. 1992; Islam 1995). The variable x_{it} is some certain choice of the pollution variable to be specified later, which is either directly related to Z in (5.24) or z in (5.25).

5.4 Discussion of Some Empirical Results

In order to carrying out relevant empirical analysis to test our theoretical modeling above, we choose a sample of a panel data structure that includes all the provincial-level regions in the mainland of China over time. Except for the variable x_{it}, data on all the other variables in the regression model above can be obtained in a straightforward way from various official sources of the National Bureau of

Statistics of China. However, exact annual data on provincial (per capita) pollution emission are hard to come by directly, as pollution itself involves the emission of a wide variety of different types of pollutants. We can therefore use the emission of a certain type of key pollutant to as an index (indicator) for general pollution emission. For example, we can use the greenhouse gas emission such as carbon emission as an indicator for general pollution emission. It follows that a feasible measure of annual per capita carbon emission can be constructed in this way

$$x_{it} = \sum_j \frac{E_{jit}}{E_{it}} \cdot \frac{x_{jit}}{E_{jit}} \cdot E_{it} = \sum_j S_{jit} \cdot F_{jit} \cdot E_{it} \tag{5.28}$$

where x_{it} stands for per capita carbon emission, E_{it} for total energy resource consumption, x_{jit} for per capita carbon emission from the consumption of the jth type of energy resource, and E_{jit} for consumption of the jth type of energy resource. Therefore, S_{jit} is the share of consumption of the jth type of energy resource in total energy resource consumption, and F_{jit} is the emission coefficient of the jth type of energy resource with respect to per capita carbon emission. To keep things as simple as possible, we first use three types of energy resources, that is, petroleum, coal and natural gas, to formulate the measure of carbon emission in (5.28). Data needed for this computation can be found in official sources of the National Bureau of Statistics of China.

We use several different data setups in our regression analysis, where each period t in (5.27) can pertains to any feasible length of calendar years, so that with any certain span of the sample period (usually from the early 1990s to the very recent years) we can run through the whole sample period continuously by taking out the series of sub-periods of time spans of the pre-set length as the time periods i. Therefore, for any given sample span and our chosen time periods i, we design the corresponding set of time dummies, along with a common intercept, to take account of the time intercept η_t in (5.27).

One of our major goals is to estimate the structural parameters in the model, which are, namely, α and γ. In addition to estimating the potential values of the structural parameters α and γ, we also aim to estimate the magnitude of the speed of convergence λ. However, our primary interest is in the magnitude of the coefficient on (partial effect of) the pollution emission term in (5.27). Our estimation results generally suggest that the estimated coefficients on the explanatory variables are statistically significant (at the 5 % level) and have the expected signs.[4] The estimated convergence speed λ is around 0.05–0.10. This indicates that controlling for the other variables (including the fixed region effects), per capita output converges to its steady state value at an annual speed of about 5–10 %. The estimated values of α turns out to be about 0.35–0.60, which fall within the likely range of its theoretically and empirically accepted values (also see Chap. 3). The estimated

[4]Because there are many regressions with alternative variable formations, thus to page space, we choose not to present the results explicitly.

value of γ, which is the focus of our primary interest, turns out to be in the range of 0.32–0.53. Using slightly different presumed values of the unknown parameter ξ to rerun the regression above as a robustness check, we end up with the conclusion that our estimation results are not sensitive to the presumed value of ξ.

5.5 Concluding Remarks

It has been long argued that China's economic growth in the recent decades has been relying too much heavily on the extensive use of environmental inputs. In this chapter, based on the idea of a theoretical framework and econometric modeling, we have empirically analyzed the contribution of pollution emission (using various pollutant emissions as the representative pollution emission) as an environmental input to the growth of China's income. We build our empirical discussion on a theoretical framework of environmentally sustainable growth. Based our theoretical discussion, we have reached the result that for environmentally friendly growth to be sustainable, we need to make the values of g and b sufficiently large in the condition (5.21). This implies that, given the values of g and b, a large ratio of γ to β is evidence against environmentally sustainable growth. Our regression exercise, however, has generally produced estimates of γ high enough compared with corresponding estimates of β. Therefore, our regression results, though still very rough, give rise to concerns about the sustainability of China's economic growth. We reasonably suspect that environmentally friendly economic growth is not being sustained in China. Relevant policy measures regarding pollution emissions and pollution abatement may therefore be urgently needed.

References

Abdulai, Awudu, and Linda Ramcke. 2009. The Impact of Trade and Economic Growth on the Environment: Revisiting the Cross-Country Evidence. Kiel Working Paper No. 1491, Kiel Institute for the World Economy, March 2009.

Antweiler, W., B.R. Copeland, and M.S. Taylor. 2001. Is free trade good for the environment? *American Economic Review* 91(4): 877–908.

Arrow, Kenneth J., Partha Dasgupta, Lawrence H. Goulder, Kevin J. Mumford, and Kirsten Oleson. 2012. Sustainability and the measurement of wealth. *Environment and Development Economics* 17(3): 317–353.

Becker, R. 1982. Intergenerational equity: The capital environment trade-off. *Journal of Environmental Economics and Management* 9: 165–185.

Bovenberg, A.L., and S. Smulders. 1995. Environmental quality and pollution augmenting technological change in a two sector endogenous growth model. *Journal of Public Economics* 57: 369–391.

Brock, William A. 1973. Polluted Golden Age. In *Economics of Natural and Environmental Resources*, ed. V.L. Smith. New York: Gordon & Breach.

Brock, William A, and M. Scott Taylor. 2004. The Green Solow Model. NBER Working Paper No. 10557.

Brock, W.A., and M. Scott Taylor. 2005. Economic growth and the environment: A review of theory and empirics. In *The handbook of economic growth*, eds. P. Aghion and S. Durlauf. A volume in the series Handbooks in Economics, edited by, Kenneth Arrow and Michael D. Intrilligator. Elsevier Publisher.

Co, C.Y., F. Kong, and S. Lin. 2008. Pollution across Chinese Provinces. Development Economics Working Papers 22148, East Asian Bureau of Economic Research.

Considine, T.J., and D.F. Larson. 2006. The environment as a factor of production. *Journal of Environmental Economics and Management* 52(3): 645–662.

Copeland, B.R., and M.S. Taylor. 1994. North-South trade and the global environment. *Quarterly Journal of Economics* 109: 755–787.

Gale, L.R., and J.A. Mendez. 1998. The empirical relationship between trade, growth and the environment. *International Review of Economics and Finance* 7(1): 53–61.

Grossman, G.M., and A.B. Krueger. 1993. Environmental impacts of the North American free trade agreement. In *The U.S.-Mexico Free Trade Agreement*, ed. P. Garber, 13–56. Cambridge, MA: Massachusetts Institute of Technology, The MIT Press.

Grossman, G.M., and A.B. Krueger. 1995. Economic growth and the environment. *Quarterly Journal of Economics* 110(2): 353–377.

Islam, Nazrul. 1995. Growth empirics: A panel data approach. *Quarterly Journal of Economics* 110: 1127–1170.

Jiang, Yanqing. 2014. Openness, economic development, and the environment in Post-Reform China: A model with empirical analysis. *International Journal of Social Science Studies* 2(3): 56–66.

Jiang, Yanqing. 2015. Is environmentally sustainable growth being achieved in China? An analysis of pollution emission as a production input. *Scholars Journal of Economics, Business and Management* 2(2A): 159–164.

Levinson, A. 2002. The ups and downs of the environmental Kuznets curve. In *Recent advances in environmental economics*, ed. J. List, and A. de Zeeuw. Cheltenham: Edgar Elgar.

Mankiw, N.G., D. Romer, and D.N. Weil. 1992. A contribution to the empirics of economic growth. *Quarterly Journal of Economics* 107: 407–437.

Mohtadi, H. 1996. Environment, growth and optimal policy design. *Journal of Public Economics* 63: 119–140.

Nordhaus, W.D., and J. Boyer. 2000. *Warming the world: Economic models of global warming*. Cambridge, MA: MIT Press.

Selden, T.M., and D. Song. 1994. Environmental quality and development: Is there a Kuznets curve for air pollution emissions? *Journal of Environmental Economics and management* 27: 147–162.

Smulders, S., and R. Gradus. 1996. Pollution abatement and long term growth. *European Journal of Political Economy* 12: 505–532.

Solow, Robert. 2012. A few comments on 'Sustainability and the measurement of wealth'. *Environment and Development Economics* 17(3): 354–355.

Tahvonen, O., and J. Kuuluvainen. 1993. Economic growth, pollution and renewable resources. *Journal of Environmental Economics and Management* 24: 101–118.

Tol, R.S.J. 2009. The economic effects of climate change. *Journal of Economic Perspectives* 23(2): 29–51.

Tzouvelekas, E., D. Vouvaki, and A. Xepapadeas. 2006. Total Factor Productivity Growth and the Environment: A Case for Green Growth Accounting. Working Papers 0617, University of Crete, Department of Economics.

Vincent, J.R. 1997. Testing for environmental Kuznets curves within a developing country. *Environment and Development Economics* 2: 417–431.

World Bank. 2006. *Where is the wealth of nations? Measuring capital for the 21st century*. Washington, DC: World Bank.

World Bank. 2011. *The changing wealth of nations: Measuring sustainable development in the new millennium*. Washington, DC: World Bank.

Xepapadeas, A. 2003. Economic growth and the environment. In *Handbook of environmental economics*, eds. Karl Göran Mäler and Jeffrey Vincent.
Xepapadeas, A. 2005. Economic Growth and the environment. In *The Handbook of Environmental Economics, Economy wide and International Environmental Issues*, eds. Mäler, K.G. and J. Vincent. A volume in the series Handbooks in Economics, edited by, Kenneth Arrow and Michael D. Intrilligator. Amsterdam: Elsevier Publishers.
Zhang, Qingfeng, and Robert Crooks. 2011. Environmental Strategy for the 12th Five-Year Plan Period: What Can the People's Republic of China Learn from the 11th Five-Year Plan? *ADB Briefs* No 8, June 2011.

Chapter 6
Foreign Trade and Sustainable Development

6.1 Introduction

In this chapter we focus on the relationship between foreign trade and sustainable development in the case of the Chinese regions. Foreign trade may affect pollution outcomes and the environment via various different channels. The first channel is the so-called scale effect that is usually thought of as harmful to the environment, which pertains to the scenario where the expansion of domestic production pressured by increasing trade openness results in more environmental pollution. Another channel is the technology effect which is considered conducive to the environment, that is, domestic firms are able to benefit from the spillovers induced by foreign trade and thus have access to the use of more advanced technologies that are less pollutive. Through this channel, trade openness can potentially improve the environmental quality by making firms less pollutive. Still another channel is the composition effect, which is related to the structural transformation that occurs in the economy driven by foreign trade, that is, foreign trade shapes the pattern of production specialization among different countries, and thus by reallocating resources across different domestic sectors, affects the overall pollution intensity of the various production activities within the country. One more channel is the income channel. Together with income gains introduced by foreign trade, the country may desire higher environmental standards, more stringent regulations, and better enforcement by the government, which are all beneficial to the environment (See Jiang 2014a).

As mentioned in earlier chapters, the EKC framework mainly focused on the effect of economic development on environmental degradation. However, the pollution consequences of economic growth depend on the underlying sources of growth (see again, for example, Antweiler et al. 2001). The EKC framework does not usually take account of the effects of foreign trade on the environment. But in fact, as mentioned above, trade patterns may (partly) explain reductions in pollution in high-income countries and increases in pollution in low-income countries. A case

Y. Jiang, *Green Development in China*,
SpringerBriefs in Economics, DOI 10.1007/978-981-10-0693-7_6

in point is that the pollution haven hypothesis actually proposes that disparities in the stringency of environmental regulations between high-income and low-income countries will generally lead to increased pollution-intensive production in the developing countries (Cole 2004). In comparison, another hypothesis, the factor endowment hypothesis, maintains that according to trade and specialization patterns, countries that are relatively abundant in factors used intensively in polluting industries will generate more pollution as trade barriers are lifted (Mani and Wheeler 1998).

We are therefore interested in uncovering the environmental implications of China's processes of opening up and economic development. In this chapter, based on some theoretical framework, we will empirically examine the effects of trade openness and the level of development on pollution emission across the Chinese provinces. This chapter is organized as follows. In Sect. 6.2, we present our basic theoretical framework. In Sect. 6.3, we discuss issues related to the empirical setup, including the sample, data, variables and the regression specifications. In Sect. 6.4, we present and discuss our regression results. Section 6.5 concludes this chapter.

6.2 The Theoretical Foundation

We first recapitulate what we manifested in Chap. 3 about sustainability. Following Arrow et al. (2012) definitions and notations (see also Chap. 3 earlier), we can show that intergenerational well-being (at time t) can be expressed as

$$V(t) = \int_{t}^{\infty} [U(\mathbf{C}(s))e^{-\delta(s-t)}ds] \tag{6.1}$$

where $\delta > 0$ is the felicity discount rate. An economic forecast at time t is the vectors $\{\mathbf{C}(s), \mathbf{K}(s)\}$ for $s > t$. We can now define that economic development (growth) is sustained at time t is if $dV/dt \geq 0$.[1] Thus, given $\mathbf{K}(t)$, $\mathbf{K}(s)$ and $\mathbf{C}(s)$ and therefore $U(\mathbf{C}(s))$ are determined for all future times $s > t$. We can then write

$$V(t) = V(\mathbf{K}(t), t) \tag{6.2}$$

where V also depends directly on t.

With such a definition of sustainability, we move on to a model about foreign trade and its effect on environmental quality. We follow Copeland and Taylor (2004) in presenting the simplest framework of the model. We assume the economy produces two goods, X and W, each with a constant returns to scale technology with respect to two inputs, labor and capital. The production of X generates

[1] We need to assume that the integral in (6.1) converges for the forecast.

pollution but the production of W does not. Denoting the price of good X by p with good W as the numeraire, we have the production of good W as

$$w = H(K_w, L_w) \tag{6.3}$$

and the production of good X as

$$x = z^{\alpha}[F(K_x, L_x)]^{1-\alpha} \tag{6.4}$$

in which z is pollution emission.[2] If the firms face a pollution tax τ for each unit of pollution they emit, then with a Cobb-Douglas functional form, we have

$$z/x = \alpha p/\tau \tag{6.5}$$

where z/x is the emission intensity of good X. We see that emission is inversely related to the pollution tax τ, and positively related to the price of good X, p.[3]

Therefore, The production of both goods can be written as functions of factor endowments, prices and the environmental policy

$$w = w(p, \tau, K, L) \text{ and } x = x(p, \tau, K, L) \tag{6.6}$$

National income, G, for any given level of pollution emission z, can be written as

$$G(p, K, L, z) = \max\{px + w\colon (x, w) \in T(K, L, z)\} \tag{6.7}$$

where T is the feasible technology set. It can be shown that $\tau = \partial G/\partial z$. Assume N identical consumers in the economy with each maximize her utility, taking pollution as given. The indirect utility function of one representative consumer can be expressed as

$$V(p, I, z) = v(I/\beta(p)) - h(z) \tag{6.8}$$

where $I \equiv G/N$ is the per capita income, β is a price index, and h is increasing and convex while v is increasing and concave.

The government chooses the pollution level so that utility of the representative consumer is maximized subject to production possibilities and private sector behavior. The government's problem is

$$\max_{z} V(I/\beta(p), z), \quad \text{s.t. } I = G(p, K, L, z)/N \tag{6.9}$$

---

[2]Both production functions are increasing, concave and linearly homogeneous.

[3]For convenience, we assume that the economy is small in the world market and the government takes the price of good X as given.

for which the first-order condition is

$$V_I G_z / N + V_z = 0 \tag{6.10}$$

With $G_z = \tau$ (6.10) can be rearranged into

$$\tau = G_z(p, K, L, z) = N \cdot (-V_z / V_I) = N \cdot \vartheta(p, R(p, K, L, z), z) \tag{6.11}$$

where we define $R \equiv I/\beta(p)$ as the real income of the typical consumer, and $\vartheta(p, R, z)$ is the typical consumer's marginal damage from pollution. Equation (6.11) determines the efficient level of pollution z^*. To implement z^*, the government can either introduce the accompanying pollution tax τ^*.[4]

6.3 The Empirical Setup and Data

The important point we gain from the theoretical model above is that the economy confronts a tradeoff between goods production and environmental quality, given the constraint of a certain level of technology. Motivated by the theoretical model, we are interested in empirically investigating the possible effects of foreign trade on pollution emission. Considering the EKC framework we discussed in previous chapters and employing a panel data structure as in Jiang (2015), we can write our baseline regression specification as

$$\begin{aligned} \ln m_{it} = {} & \beta_{T1} \ln T_{it} + \beta_{T2} (\ln T_{it})^2 + \beta_{y1} \ln y_{it} + \beta_{y2} (\ln y_{it})^2 + \beta_{k1} \ln k_{it} + \beta_{k2} (\ln k_{it})^2 \\ & + \beta_{h1} \ln h_{it} + \beta_{h2} (\ln h_{it})^2 + \beta_{L1} \ln L_{it} + \beta_{L2} (\ln L_{it})^2 + \tau_t + \varphi_i + \varepsilon_{it} \end{aligned} \tag{6.12}$$

where the dependent variable m_{it} is some measure of regional per capita pollution emission, and the independent variables are our measures of regional trade openness T_{it}, regional per capita income (i.e. per capita GDP) y_{it}, regional per capita physical capital stock k_{it}, regional per capita human capital stock h_{it}, and regional population L_{it}.[5]

[4] Or alternatively the government may issue the corresponding marketable pollution quotas that would yield the equilibrium level of pollution tax.

[5] The squared terms of the relevant independent variables are also included in the equation in order to take account of potential nonlinear partial effects of these variables as predicted by the EKC framework. The last three terms are the time varying intercept, the time constant regional heterogeneity, and the zero-mean idiosyncratic error term, respectively.

Our various sample choices are the same as in the previous chapter. Exact annual data on regional carbon emission (defined $M \equiv mL$) are hard to obtain directly. However, an effective measure of annual regional carbon emission can be constructed in the following way (using the same method as in Eq. (5.28) in Chap. 5)

$$M_{it} = \sum_{j} \frac{E_{jit}}{E_{it}} \cdot \frac{M_{jit}}{E_{jit}} \cdot E_{it} = \sum_{j} S_{jit} \cdot F_{jit} \cdot E_{it} \tag{6.13}$$

where M_{it}, as defined above already, is total regional carbon emission (of province i in period t), E_{it} is total regional energy consumption, M_{jit} is regional carbon emission from the consumption of the jth type of energy, and E_{jit} is regional consumption of the jth type of energy. Therefore, S_{jit} denotes the share of regional consumption of the jth type of energy in total regional energy consumption, and F_{jit} denotes the emission intensity (coefficient) of the jth type of energy regarding carbon emission.[6] Once data on M_{it} are obtained, it is then straightforward to obtain data on regional per capita carbon emission m_{it}.

The trade openness variable T_{it} is normally constructed as the share of regional foreign trade in regional income, where relevant data are also available from official publications of the National Bureau of Statistics of China. In this study we make adjustment to this trade openness measure in order to take account of a broader sense of openness, by following the method of Low, Olarreaga and Suarez (1998). Likewise, data on regional population L_{it} and regional (real) per capita income y_{it} can also be directly calculated from official sources from the National Bureau of Statistics of China. Further, data on regional per capita physical and human capital stocks, k_{it} and h_{it}, can be constructed by following the method of Jiang (2014b).

As in Jiang (2015), beside the regression specification in (6.12), an alternative regression specification we would like to apply is

$$\begin{aligned} \ln m_{it} = {} & \beta_{T1} \ln T_{it} + \beta_{T2} (\ln T_{it})^2 + \beta_{A1} \ln A_{it} + \beta_{A2} (\ln A_{it})^2 + \beta_{k1} \ln k_{it} + \beta_{k2} (\ln k_{it})^2 \\ & + \beta_{h1} \ln h_{it} + \beta_{h2} (\ln h_{it})^2 + \beta_{L1} \ln L_{it} + \beta_{L2} (\ln L_{it})^2 + \tau_t + \varphi_i + \varepsilon_{it} \end{aligned} \tag{6.14}$$

where we substitute the level of TFP (denoted A_{it}) for per capita income y_{it} in (6.12). One potential advantage of (6.14) over (6.12) is that the former may involve less multicollinearity among its independent variables. The needed levels of TFP can be computed by using a method similar to that in Jiang (2014b).

[6]In this study, owing to data availability, we choose three types of energy, namely, petroleum, coal, and natural gas for constructing our measure of regional carbon emission based on the formula above. Relevant data needed for the construction can be found in various official publications of the National Bureau of Statistics of China.

6.4 The Empirical Setup and Estimation Results

Just like in Chap. 5, we try several different data setups in our regression analysis, where each period t in (6.12) and (6.14) can correspond to any feasible length of calendar years, so that with any certain length of the sample period (normally from the early 1990s to the very recent years) we can run through the whole sample period successively by sorting out the series of sub-periods of time spans of the pre-set length as the time periods i. Therefore, for any given sample length and our chosen time periods i, we design the corresponding set of time dummies, along with a common intercept, to take account of the time intercept τ_t in (6.12) and (6.14). We run various types of estimations based on (6.12) and (6.14) respectively. These estimation procedures include the pooled ordinary least squares (OLS), the fixed effects (FE), the GLS random effects (GLS RE) and the ML random effects (ML RE).

Of all the estimation procedures, our preferred estimation is the fixed effects (FE) estimation. This is because this estimation procedure controls for the time-constant region heterogeneity. We also run other estimations, namely, the GLS random effects (GLS RE), the ML random effects (ML RE) and the OLS estimators to accompany the FE estimation mainly the sake of comparison purposes.

Regarding the specification in (6.12), the FE estimation generates significant estimated coefficients on most of the explanatory variables (terms).[7] Particularly, the estimated coefficient on $\ln T_{it}$ in various similar regressions is about 0.15–0.42, which is usually significantly positive. The estimated coefficient on the squared term of $\ln T_{it}$ varies across different regressions, but is usually also positive and significant. The two estimates suggest that other things equal, that is, once the effects of regional per capita income, regional population, and regional per capita physical capital and human capital stocks (as well as the individual province heterogeneity) are accounted for, regional openness to foreign trade exerts a positive partial effect on regional per capita pollution emission. The actual mechanism for this partial effect deserves further scrutiny. Our conjecture is that regional openness to foreign trade may affect regional pollution emission by affecting the use of regional input of natural resources and/or the regional industry mix (the relative shares of the output and inputs across different sectors). A change in either of the two is necessarily connected with a change in regional TFP.

The estimated coefficient on $\ln y_{it}$ is around 2, which is usually significantly positive, depending on the specific regression, while that on the squared term, $(\ln Y_{it})^2$, is mostly significantly negative. With the effect of regional population being accounted for in the regression, this result supports the EKC hypothesis by showing that with (per capita) income increasing, pollution emission (and emission intensity) first rises and then declines.

[7]Statistic significance discussed here in this analysis is at the usual 5 % level unless otherwise stated.

The estimated coefficients on $\ln k_{it}$ and $(\ln k_{it})^2$ are both insignificant. This implies that once all the other explanatory variables are controlled for, physical capital is not shown to have a significant partial effect on pollution emission. The estimated coefficient on $\ln h_{it}$ is mostly significantly negative, while that on its squared term, $(\ln h_{it})^2$, is insignificant (insignificantly negative). This result indicates that, other factors being equal, a higher level of regional per capita human capital is associated with a lower level of regional per capita pollution emission.

The estimated coefficient on $\ln L_{it}$ is mostly significantly negative, and the one on the squared term, $(\ln L_{it})^2$ is usually significantly positive. When the effects of the other explanatory variables are partialed out, a change in regional population must necessarily be associated with a change in either regional TFP or/and the level of regional resource input. Both of the two may exert an impact on regional pollution emission.

Regarding the specification in (6.14), the FE estimation also generates significant estimated coefficients on most of the explanatory terms. Particularly, the estimated coefficients on $\ln T_{it}$ and its squared term $(\ln T_{it})^2$ are both significantly positive. These two estimates show that when the partial effects of regional TFP, regional population, and regional per capita physical capital and human capital stocks (as well as the individual province heterogeneity) are accounted for, regional openness to foreign trade is likely to have a positive partial effect on regional per capita pollution emission. The underlying mechanism for this positive partial effect again deserves further investigation. A plausible explanation is that regional openness to foreign trade may affect regional pollution emission by affecting regional input of natural resources, a change in which is in turn associated with a change in regional output.

The estimated coefficient on $\ln A_{it}$ is around 2 to 3, which is significantly positive, while that on its squared term, $(\ln A_{it})^2$, is significantly negative. Therefore, with the other explanatory variables being controlled for, this result also tends to support the EKC hypothesis by showing that as TFP, hence (per capita) income, increases, pollution emission (and emission intensity) first rises and then declines.

The estimated coefficient on $\ln k_{it}$ is mostly significantly positive but that on $(\ln k_{it})^2$ is almost always insignificant. The estimated coefficient on $\ln h_{it}$ is mostly significantly negative, while that on its squared term, $(\ln h_{it})^2$, is almost always insignificant (insignificantly negative). Therefore, ceteris paribus, our results show that a higher level of regional per capita human capital is associated with a lower level of regional per capita pollution emission.

The coefficient on $\ln L_{it}$ is estimated to be negative, and the one on its squared term, $(\ln L_{it})^2$, is estimated to be positive, both of which are very statistically significant and practically large. Again, such results suggest that when the partial effects of all the other explanatory variables are accounted for, a change in regional population should be associated with a change in the level of regional resource input, which may in turn have an impact on regional pollution emission.

6.5 Concluding Remarks

On the one hand, foreign trade is widely considered one of the key driving forces behind China's miraculous economic development in the past few decades. On the other hand, China's economic development has been relying extensively on environmental inputs. Excessive exploitation of environmental resources is threatening China's ecosystem and natural resources, making the prospect of environmentally sustainable development highly unlikely. In this chapter we focus on exploring the relationship between foreign trade and sustainable development in the case of the Chinese regions.

Our empirical results suggest that when the effects of per capita income, population, and per capita physical capital and human capital stocks are netted out, openness to foreign trade exerts a significant positive partial effect on per capita pollution emission. The results lead us to suspect that openness to foreign trade may affect pollution emission by affecting the input demand for natural resources and/or the industry mix. Our empirical results also seem to support the EKC hypothesis by showing that as (per capita) income or TFP increases pollution emission and emission intensity both first rise and then decline. We see that although foreign trade may be an engine for economic growth, it may also be a key contributor to a deteriorating environment. Therefore, when designing development strategies, the possible negative environmental impacts of foreign trade should be fully considered and gauged in order to achieve environmentally sustainable development.

References

Antweiler, W., B.R. Copeland, and M.S. Taylor. 2001. Is free trade good for the environment? *American Economic Review* 91(4): 877–908.

Arrow, Kenneth J., Partha Dasgupta, Lawrence H. Goulder, Kevin J. Mumford, and Kirsten Oleson. 2012. Sustainability and the measurement of wealth. *Environment and Development Economics* 17(3): 317–353.

Cole, M.A. 2004. Trade, the pollution haven hypothesis and the environmental kuznets curve: Examining the linkages. *Ecological Economics* 48: 71–81.

Copeland, B.R., and M.S. Taylor. 2004. Trade, growth, and the environment. *Journal of Economic Literature* 42(1): 7–71.

Jiang, Yanqing. 2014a. Openness, economic development, and the environment in post-reform China. A model with empirical analysis. *International Journal of Social Science Studies* 2(3): 56–66.

Jiang, Y. 2014b. Understanding TFP growth in Inland Regions of China: An empirical study of the effects of three factors. *International Review of Applied Economics* 28(3): 364–381.

Jiang, Yanqing. 2015. Foreign trade and carbon emission: Empirical evidence from China. *Scholars Journal of Economics, Business and Management* 2(1B): 72–79.

Low, Patrick, Marcelo Olarreaga and Javier Suarez. 1998. Does globalization cause a higher concentration of international trade and investment flow? WTO Staff Working Paper ERAD-98-08 (August 1998), Economic Research and Analysis Division, World Trade Organization.

Mani, M., and D. Wheeler. 1998. In search of pollution havens? Dirty industry in the world economy, 1960–1995. *Journal of Environment and Development* 7(3): 215–247.

Chapter 7
Foreign Direct Investment and Sustainable Development

7.1 Introduction

Continuing from our discussions in the preceding chapter, we now focus on the relationship between foreign direct investment (FDI) and sustainable development in the case of the Chinese regions in this chapter. Undoubtedly huge inflows of FDI were an important driving force behind China's economic development. FDI may exert important environmental consequences. These environmental consequences can occur via various different channels. These channels are parallel to those we discussed in Chap. 6 through which foreign trade may affects the environmental quality.

Here we draw on Jiang (2015) to briefly discuss the channels. The first channel is the technique effect that generally contributes to a better environment. Via this channel, FDI could improve the environmental quality by providing access for domestic firms to technological spillovers from foreign firms so that domestic firms can adopt more advanced technologies that are less pollutive. Another channel is the scale effect, which may be harmful to the environment. The scale effect may result in a situation where the scaling up of the economy pulled by increasing FDI leads to more environmental pollution. A third channel is the income effect. That is, with income increases brought about by FDI, the country may live up to higher environmental standards, more stringent regulations, and better enforcement by the government, which are all beneficial to the environment. Still another channel is the competition effect. FDI plays a role because countries (or regions within a country) may compete strategically to attract and retain FDI by racing down to the bottom of their environmental standards (Smarzynska and Wei 2001; King 2011; Jiang 2015). Still one more channel is the composition effect. This effect is associated with the structural change within the economy. As FDI is correlated with foreign trade of the host country and both affect the patterns of division of labor and production specialization between the country and the outside world, FDI may affects the overall

Y. Jiang, *Green Development in China*,
SpringerBriefs in Economics, DOI 10.1007/978-981-10-0693-7_7

pollution intensity in the country by shaping the reallocation of resources across different domestic sectors.[1]

The simplest EKC framework obviously did not account for the effects of FDI on the environment. But in fact, the spatial pattern of FDI flows may help explain reductions in pollution in high-income countries (regions) and increases in pollution in low-income (regions) countries. As mentioned earlier, the pollution-haven-seeking behavior of FDI implies that differences in the stringency of environmental regulations across different countries or regions will generally result in increased pollution-intensive production in the underdeveloped countries or regions.

There have been a handful of recent empirical studies on the relationship of FDI and the environment in China. For example, Zheng et al. (2010) investigated the relationship between FDI and air pollution across major Chinese cities. Their empirical results suggested that per capita FDI flows were inversely related to pollution levels across the Chinese cities. Kim and Adilov (2011) focused on the relationship between FDI and pollution represented by carbon dioxide emissions. They found that although in low-income countries slack environmental regulations may help attract FDI, the foreign firms were inclined to use technology that were less pollutive than that used by local firms. Therefore, FDI did not necessarily boost pollution levels in the host countries. Omri et al. (2014), by applying dynamic simultaneous equations panel data models, tested the causal relationship between carbon dioxide emissions, FDI, and economic growth. They found evidence of bi-directional causal relationship between FDI inflows and economic growth and between FDI and carbon dioxide emissions.

In the study of this chapter, we are interested in uncovering the environmental impacts of China's processes of economic development and opening up to FDI. Specifically, we empirically examine the potential effects of FDI and the level of economic development on pollution emissions across the Chinese regions. This chapter is structured as follows. In Sect. 7.2, we present our theoretical framework, which inspires and motivates our subsequent econometric analysis, as well as the basic econometric setup that follows the theoretical framework. In Sect. 7.3, we discuss the data and related issues in preparation of our estimation analysis. In Sect. 7.4, we discuss our estimation results. Section 7.5 concludes this chapter.

7.2 A Theoretical Framework and Econometric Setup

To keep things as simple as possible, we first consider single-sector economy for illustration purposes. At any point in time, we can consider aggregate output to be associated with the various needed production input by the aggregate production

---

[1]This chapter is closely related to the preceding one because FDI inflows and foreign trade in China were always closely linked. Foreign Invested Enterprises (FIE's) generally accounted for over 50 % of China's exports and over 60 % of China's imports (see Whalley and Xin 2010).

function whose specific functional form is not of central important for our current study, that is

$$Y_t = Y(K_t, F_t, H_t, L_t, R_t, A_t) \tag{7.1}$$

where Y_t is output, K_t is domestically invested physical capital stock, F_t is FDI capital stock, H_t is (a certain measure of) human capital stock, L_t is the amount of (raw) labor input, R_t is the quantity of natural resources (including land, clean air, clean water, etc.) used in production, and A_t is the level of technology or total factor productivity (TFP).

One important point here is that, as we further assume, pollution emission is a function of all the arguments in (7.1). That is, pollution emission does not directly correspond to the quantity of output, but to the quantities of the different production inputs, i.e. the arguments in (7.1).

Therefore pollution emission E_t can be expressed as

$$E_t = E[K_t, F_t, H_t, L_t, R_t, A(K_t, F_t, H_t, L_t, R_t, Y_t)] = \tilde{E}(K_t, F_t, H_t, L_t, R_t, Y_t) \tag{7.2}$$

where the first equality comes from the fact that the level of TFP, A_t, is not directly observable and can thus be written (implicitly) as a function of the other variables according to the production function (7.1), and the second equality then redefines the function and denotes it with a tilde overhead. Equation (7.2) shows that at any time t, pollution emission is determined jointly by six variables, which are domestically invested physical capital, FDI capital, human capital, labor, the quantity of natural resources used in production and output.

The modeling above can be extended to a multi-sector situation. Consider an economy of m sectors. Pollution emission of sector j (j = 1, 2, …, m) can be written as

$$E_t^j = \tilde{E}^j(\theta_t^{Kj} K_t, \theta_t^{Fj} F_t, \theta_t^{Hj} H_t, \theta_t^{Lj} L_t, \theta_t^{Rj} R_t, \theta_t^{Yj} Y_t) \tag{7.3}$$

where the θ's are the shares of economy-wide K_t, F_t, H_t, L_t, R_t and Y_t in sector j, with all the corresponding θ's summing up to one. Total pollution emission of the entire economy can then be rewritten as

$$E_t = \tilde{\tilde{E}}(K_t, F_t, H_t, L_t, R_t, Y_t, \Theta_t) \tag{7.4}$$

where Θ_t is a vector collecting all the θ's.

Based on the idea presented above, we can formulate our econometric model, which may adopts a panel data structure as follows

$$\begin{aligned}\ln E_{it} = {} & \beta_{F1} \ln F_{it} + \beta_{F2} (\ln F_{it})^2 + \beta_{Y1} \ln Y_{it} + \beta_{Y2} (\ln Y_{it})^2 + \beta_{K1} \ln K_{it} + \beta_{K2} (\ln K_{it})^2 \\ & + \beta_{H1} \ln H_{it} + \beta_{H2} (\ln H_{it})^2 + \beta_{L1} \ln L_{it} + \beta_{L2} (\ln L_{it})^2 + \tau_t + \varphi_i + \varepsilon_{it}\end{aligned} \tag{7.5}$$

The dependent variable E_{it} is regional pollution emission (of region i in year t). The independent variables are regional FDI capital stock F_{it}, regional output Y_{it}, regional domestically invested physical capital stock K_{it}, regional human capital stock H_{it}, and regional labor input L_{it}. In Eq. (7.5), total regional levels are used instead of regional per capita (per worker) levels. This will not necessarily lead to bias problems as the variables are entered in logs and labor is also explicitly entered into the regression equation. To account for potential nonlinearity in partial effects of the variables as captured by the EKC framework, we include in the regression equation the squared terms the variables. The variable associated with natural resources is dropped from the regression specification for the reason of data unavailability.

In Eq. (7.5), the FDI variable F_{it} is of our primary interest. The partial effect of $\ln F_{it}$ on $\ln E_{it}$ is determined by $\beta_{F1} + 2\beta_{F2} \ln F_{it}$, where a non-zero value of β_{F2} indicates a nonlinear partial effect of $\ln F_{it}$ on $\ln E_{it}$. This partial effect of $\ln F_{it}$ is the effect of regional FDI on regional pollution emission when the effects of the other variables are netted out. Given the specification in (7.5), the partial effect of $\ln F_{it}$ on $\ln E_{it}$ also incorporates the effect of FDI on pollution emission via its effects on the industry mix or the input of natural resources (while a change in either one of both the industry mix or the input of natural resources must pertain to a change in TFP, as output and the other inputs are controlled for in the regression specification).[2]

The modeling above shows that when the effects of output, domestically invested physical capital, human capital and labor input are netted out in the regression, FDI may affect pollution emission via several different mechanisms. One channel may be that FDI affects pollution emission by affecting the input of natural resources.[3] This is equivalent to saying that FDI affects pollution emission by affecting TFP. This mechanism largely corresponds to the 'technique effect' mentioned above. Another mechanism may be that FDI affects pollution emission by affecting the industry mix, a change of which would generally lead to a change in economy-wide TFP. This second mechanism largely pertains to the 'composition effect' mentioned above. A third mechanism may be that FDI affects pollution emission even without affecting the input of natural resources, the industry mix, or TFP. This mechanism is primarily associated to the 'scale effect' discussed earlier. Finally, all of the three mechanisms may be related to the 'income effect' discussed earlier.

7.3 Data and Related Issues

We use variant samples to carry out our econometric analysis. Our samples include the provincial-level regions (provinces for short sometimes) in mainland China over various time periods that usually run from the late 1990s till very recent years.

---

[2]We should thus note that our regression specification in (7.5) does not explicitly control for the effects of the industry mix and the input of natural resources on pollution emission.

[3]Suppose for the moment the industry mix is held fixed.

Exact measures of regional pollution emission are quite hard to obtain. Therefore, the emissions of a few chosen pollutants are used to represent overall regional pollution emission. Just the same as in the study of Chap. 4, these pollutants are, namely, sulfur dioxide, nitrogen oxides, and smoke and dust. Relevant data on provincial emissions of these kinds of pollutants over the chosen sample periods can be obtained from the official publications of the National Bureau of Statistics of China. The publications of the National Bureau of Statistics of China also provide data on nominal GDP and GDP indices, as well as the numbers of employed people (i.e. workers), which we can use to construct the levels of regional real income and labor input.

Data on domestic and foreign investments are also obtained from the publications of the National Bureau of Statistics of China. However, these publications do not directly provide capital stock data. For this, we use a perpetual inventory method to compute domestic and FDI capital stock levels, using a 9.6 % annual depreciation rate for both (see also Zhang 2008).

As labor input is one of the explanatory variable in the regression, which is practically measured by the number of workers, it is correct to take H_{it} as a per worker measure of human capital. We follow Hall and Jones (1999) to assume that H_{it} is related to educational attainment, where the returns to schooling follow those estimated in a Mincerian wage regression (Mincer 1974; Psacharopoulos 1994). Data needed to calculate schooling can be found in the official publications of the National Bureau of Statistics of China, too.

7.4 Estimation Results

Again, we try several different data setups in our regression analysis, where each period t in (7.5) can correspond to any feasible length of calendar years, so that with any given length of the sample period (normally from the early 1990s to the very recent years) we can run through the whole sample period successively to pick out our series of sub-periods of time spans of the pre-set length as the time periods i. For any specific setup, we apply the right number of time dummies, along with a common intercept, to account for the time-variant intercept in the regression specification. We run various types of estimations based on the regression specification in (7.5). These estimation procedures include the pooled ordinary least squares (OLS), the fixed effects (FE), the GLS random effects (GLS RE) and the ML random effects (ML RE).

Just like in the previous chapter, of all the estimation procedures, our preferred estimation is the fixed effects (FE) estimation. This is because this estimation procedure controls for the time-constant region heterogeneity, as we think region heterogeneity plays a very important role in explaining regional pollution emission. For comparison, we also run other estimations, namely, the GLS random effects (GLS RE), the ML random effects (ML RE) and the OLS estimators to alongside with the FE estimations.

We focus mainly on the results from the FE estimations. The FE estimations turn out to generate significant estimated coefficients on most of the key explanatory variables. Particularly, the estimated coefficient on the FDI term in various similar regressions is almost always significantly positive, though the estimated values vary across regressions. The estimated coefficient on the squared term of FDI varies substantially across, but the values in most cases are significantly positive. The two pieces of results suggest that other factors being controlled for, regional FDI inflows turn out to have a positive partial effect on regional pollution emission. The possible reason behind this, as we conjecture, may be that regional FDI affects regional pollution emission by helping reallocate regional input of natural resources and/or altering the regional industry mix. A change in the allocation of regional input of natural resources or the shift of the regional industry mix is necessarily associated with a change in regional TFP.

The estimated coefficient on the income term is also significantly positive in most regressions, whose values vary according to the specific regression. The corresponding squared term is mostly significantly negative. Therefore, with the effect of regional population (labor force) being netted out in the regression, this result lends support to the EKC framework by suggesting that as income increases, pollution emission first rises and then declines. The estimated coefficients on physical capital and its squared term are both insignificant. This implies that with the other effects being partialed out, regional physical capital is not shown to have a significant partial effect on regional pollution emission. The estimated coefficient on the human capital term is significantly negative in most cases while that on its squared term is usually insignificant. This piece of result suggests that, with the effects of other factors being partialed out, regional human capital is negatively related to regional pollution emission.

The estimated coefficient on the labor term is always significantly negative, and the one on its squared term is always significantly positive. This means that holding the other explanatory variables fixed, a change in regional population (employment) must necessarily be connected to a change in either regional TFP or/and the level of regional resource input, both of which may possibly exert an impact on regional pollution emission.

7.5 Concluding Remarks

Just like openness to foreign trade, FDI inflows are also widely thought of as one of the major contributing forces propelling China's rapid economic transition and development in the past few decades. On the other hand, China faced huge environmental problems as its economic development was progressing. The prospect of green development is thus highly questionable. In this chapter we focus on exploring the relationship between FDI inflows and sustainable development in the case of the Chinese regions.

Through our empirical analysis, we have found that when the effects of income, population, physical and human capital are accounted for, regional FDI tends to have a significant positive effect on regional pollution emission. We thus conjecture that regional FDI inflows may affect pollution emission by altering the regional input use for natural resources and/or the regional industry mix. Like the previous chapter, the study in this chapter also provides support to the EKC framework by suggesting that with an increase in regional income or TFP, regional pollution emission tends to first rise and then decline. Therefore, being a propeller for economic transition and development, FDI may also serve as a major contributing factor to pollution and a worsening environment.

References

Hall, R.E., and C.I. Jones. 1999. Why do some countries produce so much more output per worker than others? *Quarterly Journal of Economics* 114: 83–116

Jiang, Yanqing. 2015. Foreign direct investment, pollution, and the environmental quality: A model with empirical evidence from the Chinese regions. *The International Trade Journal* 29 (3): 212–227.

Kim, Myeong Hwan, and Nodir Adilov. 2011. The lesser of two evils: An empirical investigation of fdi-pollution tradeoff. *Applied Economics* 44(20): 2597–2606.

King, John. 2011. Foreign direct investment and pollution havens. *Journal of Economics and Econometrics* 54(1): 39–47.

Mincer, Jacob. 1974. *Schooling, experience, and earnings*. New York: Columbia University Press.

Omri, A., D.K. Nguyen, and C. Rault. 2014. Causal interactions between CO_2 emissions, FDI, and economic growth: Evidence from dynamic simultaneous-equation models. *Economic Modelling* 42: 382–389.

Psacharopoulos, George. 1994. Returns to investment in education: A global update. *World Development* 22: 1325–1343.

Smarzynska, B.K., and S.J. Wei. 2001. *Pollution Havens and Foreign Direct Investment: Dirty Secret or Popular Myth*? NBER Working Papers 8465, National Bureau of Economic Research, Inc.

Whalley, John, and Xian Xin. 2010. China's FDI and Non-FDI economies and the sustainability of future high Chinese growth. *China Economic Review* 21(1): 123–135.

Zhang, Jun. 2008. Estimation of China's provincial capital stock (1952–2004) with applications. *Journal of Chinese Economic and Business Studies* 6(2): 177–196.

Zheng, S., M.E. Kahn, and H. Liu. 2010. Towards a system of open cities in China: Home prices, FDI flows and air quality in 35 major cities. *Regional Science and Urban Economics* 40(1): 1–10.

Chapter 8
Regional Disparity and Interaction in Environmental Effort

8.1 Introduction

In our discussions about economic development and the environment in earlier chapters, we took each part of China (i.e. the Chinese regions) as isolated geographical regions and we have not bothered to consider the potential spatial interactive behavior among the different regions regarding pollution emission and pollution abatement efforts. In this chapter, we turn to the consideration of the possibility of spatial strategic interaction in pollution abatement efforts across the Chinese regions, as well as its potential effects on the environmental quality and the prospect of green economic and social development.

In fact, the level of local pollution emission is obviously dependent on the level of pollution abatement effort of the local government. It is reasonable to think that the level of local pollution emission is negatively related to the level of pollution abatement effort of the local government. However, local governments in China often have little incentive to favor greater effort in environmental protection and thus in pollution abatement. Local governments tend to suffer from budget shortage and have relied heavily on transfers from the central government. Local governments in China have strong incentives to increase local output in order to increase revenue transfers from the central government, but usually have little incentive to protect the environment (Co et al. 2008). Therefore, the Chinese local regions may behave strategically among themselves in deciding on their own levels of pollution abatement efforts.

Several reasons may exist why different regions in China may interact strategically in deciding on their own level of environmental protection efforts. The first reason stems from the motivation to compete for mobile resources. For example, different regions may behave strategically to compete for foreign and domestic investment, which is often pollution-intensive, by racing down to the bottom of their environmental standards (Smarzynska and Wei 2001; King 2011). Second, potential spatial spillovers of the benefits from environmental protection and

Y. Jiang, *Green Development in China*,
SpringerBriefs in Economics, DOI 10.1007/978-981-10-0693-7_8

pollution abatement may play a role. For example, a region may choose its level of effort in pollution abatement in response to pollution abatement efforts of other regions, taking account of the spatial spillovers of the pollution abatement efforts of other regions (Brueckner 2003). Third, spatial strategic behavior in pollution abatement effort may also occur because of 'information spillover' (Besley and Case 1995). For example, residents of a region may compare and evaluate the performance of their local government in environmental protection against other regions, which may make the local government feel pressured and mimic the environmental policy of a neighboring region.

In this chapter, we discuss regional disparity and interaction in environmental protection (pollution abatement) effort. The rest of this chapter is organized as follows. In Sect. 8.2, we present a brief description of the basic theoretical framework we will use. In Sect. 8.3, we discuss various issues concerning the econometric methods, as well as the variables in preparation of our empirical analysis coming next. In Sect. 8.4, we discuss our preliminary empirical results. Section 8.5 presents more empirical analysis and results. Finally, Sect. 8.6 concludes this chapter.

8.2 A Theoretical Framework

In this section, we draw on Brueckner (2003) in presenting the theoretical framework.[1] The first type of strategic interaction framework can be referred to as a 'resource flow model' or a 'competition model'. In this model, each jurisdiction i chooses the level of a decision variable z_i. Jurisdiction i is not affected directly by the z levels in other jurisdictions, denoted by z_{-i}. But the jurisdiction is affected by the amount of a particular 'resource' that resides within its borders. Jurisdiction i's objective function is written as $\tilde{V}(z_i, s_i; X_i)$, where the level of the 'resource' in jurisdiction i. The resource is mobile across different jurisdictions. X_i is a vector of jurisdiction-specific characteristics, which help determine preferences. The distribution of the resource across jurisdictions is affected by the z choices of all jurisdictions. Therefore, the level of the resource available to jurisdiction i is given by $s_i = s(z_i, z_{-i}; X_i)$. This is to say that $\tilde{V}$ is indirectly affected by z_{-i}. Since X_i can be measured relative to the average characteristics of all jurisdictions, X_{-i} need not appear in the equation. Combing the two equations, we obtain

[1]Abundant historical literature includes Williams (1966), Pauly (1970), Oates (1972), and Boskin (1973), Anselin (1988), Kelejian and Prucha (1998), Ladd (1992), Besley and Case (1995), Shroder (1996), Wilson (1996, 1999), Murdoch et al. (1997), Smith (1997), Bivand and Szymanski (1997, 2000), Brett and Pinkse (1997, 2000), Heyndels and Vuchelen (1998), Revelli (2001, 2002), Brueckner (2000, 2003), Saavedra (2000), Buettner (2001), Hayashi and Boadway (2001), Brueckner and Saavedra (2001), Fredriksson and Millimet (2002), and Edmark (2007).

$$\tilde{V}(z_i, s_i; X_i) = \tilde{V}(z_i, s(z_i, z_{-i}; X_i); X_i) = V(z_i, z_{-i}; X_i) \tag{8.1}$$

which shows that the optimal level of z_i is an implicit function of z_{-i}. The solution to (8.1) reflects a Nash equilibrium in which jurisdiction i chooses z_i that maximizes its objective function V given z_{-i}, which is meanwhile the best choice for its competing jurisdictions given z_i. The solution can be described by a reaction function

$$z_i = h(z_{-i}; X_i) \tag{8.2}$$

The slope of the reaction function will be identically zero in the case where inter-jurisdictional competition for the resource is absent, and a test for the null hypothesis that the slope of the reaction function is zero is effectively a test for the existence of inter-jurisdictional competition.

The second type of strategic interaction framework can be generally referred to as a 'spillover model'. In this model, each jurisdiction i chooses the level of a decision variable z_i, but the jurisdiction is also directly affected by z_{-i} chosen in other jurisdictions, indicating the presence of spillovers. Thus jurisdiction i's objective function is written as $V(z_i, z_{-i}; X_i)$. Jurisdiction i choose z_i to maximize this objective function given X_i and z_{-i}. Therefore, the reaction function has the form $z_i = h(z_{-i}; X_i)$, which is exactly the same form of reaction function as Eq. (8.2) above.

In this chapter, we focus on potential strategic interaction in environmental protection efforts among the Chinese local governments. The variable z_i in such a case is the local decision variable measuring the level of the government's environmental protection effort in region i. As mentioned earlier already, there exist at least three incentives for the regions to interact strategically interaction in this case.

8.3 The Econometric Setup

Now taking the possibility of spatial strategic interaction into consideration, we improve our models in earlier chapters and formulate the following baseline specification

$$\ln E_{it} = \rho \sum_{j \neq i} w_{ij} \ln E_{j,t-1} + \mathbf{X}_{it}\boldsymbol{\beta} + \delta_t + c_i + \varepsilon_{it} \tag{8.3}$$

where $E_{j,t-1}$ refers to pollution emission in any other region j ($j \neq i$) in the preceding period $t-1$. We lag the levels of pollution emissions in the other regions for one period in an attempt to mitigate the endogeneity problem induced by simultaneity. Besides this reason, we think that it is also reasonable to assume that strategic reaction behavior occurs with a time lag. The vector $\mathbf{X}_{it}$ in (8.3) may contain all the key explanatory variables we discussed in earlier chapters. The w_{ij}'s are nonnegative weights reflecting the relevant importance of other regions j in the process of spatial strategic interaction. We can reasonably assume that the relevant importance is positively related to the geographical proximity between regions i and j.

Different weighting schemes for the w_{ij}'s in (8.3) would reflect different spatial patterns of strategic interaction among the regions. Under a 'contiguity scheme', we let $w_{ij} = 1$ for region j that borders on region i, and $w_{ij} = 0$ otherwise. For each i, the assigned weights are then normalized to make their sum equal unity. An alternative scheme is a 'smooth distance decay scheme', in which we assign the weights by letting $w_{ij} = 1/d_{ij}$ for $j \neq i$, where d_{ij} is the distance between the capital cities of region i and region j. The unit of measurement for the distances does not matter as for each region the assigned weights will eventually be normalized to sum up exactly to unity. Brueckner and Saavedra (2001) apply and discuss both weighting schemes. In this chapter, we follow Brueckner and Saavedra (2001) and also use both of the two weighting schemes.

8.4 Preliminary Empirical Results

We now run regressions based on (8.3). The regressions are similar to those in the last two chapters but now we include the additional explanatory variable $\sum_{j \neq i} w_{ij} \ln E_{j,t-1}$ in the regression specification. It turns out that the estimates of the coefficient on $\sum_{j \neq i} w_{ij} \ln E_{j,t-1}$ are nearly all insignificant in all of the different versions of the regressions we have come up with. Through these regressions we have not found evidence for the existence of spatial strategic interaction among the Chinese regions regarding their efforts in pollution abatement and environmental protection. In addition, we have also seen that the inclusion of the independent variable $\sum_{j \neq i} w_{ij} \ln E_{j,t-1}$ in the regression specification has not altered the estimates of all the other coefficients in any significant manner. All major results obtained from our empirical analysis in the last two chapters are not contradicted by our current study in the chapter.

8.5 Further Analysis and More Results

We suspect the chosen dependent variable, namely, (some certain measure of) regional pollution emission in Eq. (8.3) may not be a good variable that can reflect spatial strategic interaction in environmental protection or pollution abatement. Therefore, we conduct more regression exercise with what we think is a better variable associated with environmental protection effort.

According to the reaction function (8.2) earlier, we now design a z variable by dividing the environmental protection expenditure (*EPE*) of the local government by the local GDP, that is

$$z_{it} \equiv \frac{EPE_{it}}{GDP_{it}} \tag{8.4}$$

where we use provincial level data considering data availability. This ratio of provincial *EPE* to provincial GDP hopefully reflects the provincial government's effort in environmental protection. To justify that this is a proper indicator for environmental protection effort, we assume that environmental protection involves only pollution abatement. Pollution emitted equals pollution created minus pollution abated, and the level of the government's effort in environmental protection (i.e. in pollution abatement) is commensurate with pollution abated divided by pollution created. If we further assume that pollution abated is commensurate with the environmental protection expenditure and that pollution created is commensurate with GDP, we end up having that the government's effort in environmental protection is commensurate with the ratio in (8.4).

The characteristics variables $\mathbf{X}_{it}$ in the regression model should aim to capture time-varying exogenous provincial characteristics that affect z_{it}, though in theory, endogenous characteristics variables can also be included in the regression equation.[2] We choose these variables for $\mathbf{X}_{it}$: provincial GDP, the percentage of the value added of the secondary industry in GDP in the province, provincial population, general budgetary revenue of the provincial government, total developed urban area in the province, the population density (i.e. population per km^2) in the urban area of the province, the percentage of illiterate people in the population (aged 15 and above) in the province, the percentage of urban population in the province, registered urban unemployment rate, total value of foreign trade in the province, and energy consumption per unit of GDP in the province. We enter all of the variables in logs into the regression equation. In choosing the characteristics variables, we avoid choosing variables that are likely endogenous as a result of reverse causality, as we see that finding exogenous instruments for endogenous variables can be hard.[3]

Data on all the chosen variables can be obtained from official data sources in China, but the data were incomplete before 2007. Therefore, we rely on a sample period staring from the year 2007. Data on *EPE* includes the expenses on the administration of environment protection, environment monitoring and supervision, pollution control, natural ecology protection, projects of virgin forests protection, reforesting farmland, controlling the sources of dust storms, returning pastureland to grassland, returning cultivated land to grassland, energy conservation, emission reduction, and comprehensive utilization of renewable energy and resources.

---

[2]Endogenous characteristics variables can be included, but finding instruments for them can be difficult.

[3]The endogeneity issue is a hard problem: even those characteristics variables we have chosen to include in the regression equation may be argued to be endogenous too because the government's environmental protection effort, as signaled by *EPE* over *GDP*, may have varying degrees of feedback effects on these chosen characteristics variables.

For comparison purposes, we apply three estimation procedures, which are the pooled OLS, the RE (random effects), the FE (fixed effects), and the FD (first differencing) estimations. Comparing results across the different estimation procedures will do us good in helping us gain more insight into the interrelationship among the explanatory variables and their individual partial effects on the explained variable. A group (with the proper number) of time dummy variables are applied to account for the time-variant intercept in the regression equation.

Under the 'smooth distance decay' weighting scheme (mentioned earlier), it turns out that none estimates of the reaction slope (which is of our primary interest) from the various estimation procedures is significant at the usual 5 % significance level. This result may stem from two possible reasons. First, the smooth distance decay weighting scheme may not give the proper weights to the provinces when specifying the relevance of these provinces in the process of inter-provincial interaction. Second, we may have included too many irrelevant explanatory variables in the regression equation so that, owing to the multicollinearity and given the relatively small number of observations available, we have failed to estimate the reaction slope very precisely. However, regressions under the 'contiguity' weighting scheme produces estimates of the reaction slope that are all statistically significant at the 5 % significance level or even at the 1 % significance level.

We have to note that although we have applied different estimation methods, our favored method is the FE estimation. This is because we think that the assumptions underlying the FE model are the most realistic for the current situation. We believe that the dependent variable is affected by a host of region-specific factors that remain fixed (at least during our sample period). It is therefore crucial that these province-specific factors be accounted for and captured in the time-constant unobserved effects in the panel data regressions. Second, the FE method is preferred to the RE method because we believe that correlation exists between the unobserved province effect and (at least some of) the explanatory variables.

8.6 Concluding Remarks

In this chapter we explore the issue of potential strategic interaction in environmental protection efforts among the Chinese regions. Based on a theoretical framework we reach our empirical reaction function, where the sign and magnitude of the reaction slope are of our primary interest. We use two distance-based weighting schemes, the smooth decay scheme and the contiguity scheme, to specify the weights involved.

Our results from various estimation procedures, of which our favored method is the FE estimation, show that the Chinese provincial governments tend to behave strategically in deciding on the levels of their environmental protection efforts. The significantly positive estimates of the reaction slope suggest that environmental protection efforts of the provincial governments are 'strategic complements'.

We have obtained more statistically significant results from regressions under the contiguity weighting scheme than from regressions under the smooth decay weighting scheme, which is a sign for the fact that the contiguity weighting scheme works better in capturing the pattern of inter-provincial interaction in the current case. This finding suggests that direct inter-provincial strategic interaction in environmental protection efforts may occur only among contiguous provinces in China.

References

Anselin, L. 1988. *Spatial econometrics*. Dordrecht: Kluwer Academic Publishers.
Besley, T.J., and A.C. Case. 1995. Incumbent behavior: Vote seeking, tax setting and yardstick competition. *American Economic Review* 85: 25–45.
Bivand, R., and S. Szymanski. 1997. Spatial dependence through local yardstick competition: Theory and testing. *Economics Letters* 55: 257–265.
Bivand, R., and S. Szymanski. 2000. Modeling the spatial impact of the introduction of compulsory competitive tendering. *Regional Science and Urban Economics* 30: 203–219.
Boskin, M.J. 1973. Local government tax and product competition and the optimal provision of public goods. *Journal of Political Economy* 81: 203–210.
Brett, C., and J. Pinkse. 1997. Those taxes are all over the map: A test for spatial dependence of municipal tax rates in British Columbia. *International Regional Science Review* 20: 131–151.
Brett, C., and J. Pinkse. 2000. The determinants of municipal tax rates in British Columbia. *Canadian Journal of Economics* 33: 695–714.
Brueckner, Jan K. 2000. Welfare reform and the race to the bottom: Theory and evidence. *Southern Economic Journal* 66: 505–525.
Brueckner, Jan K. 2003. Strategic interaction among governments: An overview of empirical studies. *International Regional Science Review* 26(2): 175–188.
Brueckner, J.K., and L.A. Saavedra. 2001. Do local governments engage in Strategic Tax Competition? *National Tax Journal* 54: 203–229.
Buettner, T. 2001. Local business taxation and competition for capital: The choice of the tax rate. *Regional Science and Urban Economics* 31: 215–245.
Co, C.Y., F. Kong, and S. Lin. (2008). Pollution across Chinese Provinces. Development Economics Working Papers 22148, East Asian Bureau of Economic Research.
Edmark, Karin. (2007). Strategic Competition in Swedish Local Spending on Childcare, Schooling and Care for the Elderly. Working Paper Series 2007: 21, Uppsala University, Department of Economics.
Fredriksson, P.G., and D.L. Millimet. 2002. Strategic interaction and the determinants of environmental policy across US States. *Journal of Urban Economics* 51: 101–122.
Hayashi, M., and R. Boadway. 2001. An empirical analysis of intergovernmental tax interaction: The case of business income taxes in Canada. *Canadian Journal of Economics* 34: 481–503.
Heyndels, B., and J. Vuchelen. 1998. Tax mimicking among Belgian municipalities. *National Tax Journal* 51: 89–101.
Kelejian, H.H., and I.R. Prucha. 1998. A generalized spatial two-stage least squares procedure for estimating a spatial autoregressive model with autoregressive disturbances. *Journal of Real Estate Finance and Economics* 17: 99–121.
King, John. 2011. Foreign direct investment and pollution havens. *Journal of Economics and Econometrics* 54(1): 39–47.
Ladd, H.F. 1992. Mimicking of local tax burdens among neighboring counties. *Public Finance Quarterly* 20: 450–467.

Murdoch, J.C., T. Sandler, and K. Sargent. (1997) A tale of two collectives: Sulphur versus nitrogen oxide emission reduction in Europe. *Economica*, 64: 381–301.

Oates, W.E. 1972. *Fiscal federalism*. New York: Harcourt Brace.

Pauly, M.V. 1970. Optimality, 'public' goods, and local governments: A general theoretical analysis. *Journal of Political Economy* 78: 572–585.

Revelli, F. 2001. Spatial patterns in local taxation: Tax mimicking or error mimicking? *Applied Economics* 33: 1101–1107.

Revelli, F. 2002. Testing the tax mimicking versus expenditure spill-over hypothesis using English data. *Applied Economics* 34: 1723–1731.

Saavedra, L.A. 2000. A model of welfare competition with evidence from AFDC. *Journal of Urban Economics* 47: 248–279.

Shroder, M. 1996. Games the states don't play: Welfare benefits and the theory of fiscal federalism. *Review of Economics and Statistics* 77: 183–191.

Smarzynska, B.K., and S.J. Wei. (2001). Pollution Havens and Foreign Direct Investment: Dirty Secret or Popular Myth? NBER Working Papers 8465, National Bureau of Economic Research, Inc.

Smith, M.W. (1997) State Welfare Benefits: The Political Economy of Spatial Spillovers. Unpublished paper, Yale University.

Williams, A. 1966. The optimal provision of public goods in a system of local governments. *Journal of Political Economy* 74: 18–33.

Wilson, John D. 1999. Theories of tax competition. *National Tax Journal* 52: 269–304.

Wilson, John D. (1996). Capital mobility and environmental standards: Is there a theoretical basis for the race to the bottom? In *Fair trade and harmonization: Prerequisites for free trade?* eds. J. Bhagwati and R. Hundee, Vol. 1. Cambridge, MA: MIT Press.

Chapter 9
Green Development in China: Looking into the Future

This short chapter completes this book by presenting a brief discussion of prospective green development in China in the future.

The ancient Chinese philosophy advocates a unification of humanity and nature. Respect for and conformation to the laws of nature is in the very heart of the Chinese culture. The long traditional agricultural civilization of China has been relying very heavily on the weather and the environment. As the Chinese population grew larger, the ecological and environmental deficit due to the expansion of traditional agricultural production also expanded. When the new China launched its reform and opening up, it became one of the fastest growing economies and has undergone the greatest industrialization and urbanization in human history. Meanwhile, the five thousand years of the Chinese civilization have come to a point where the nation's development and transformation have been achieved at tremendous ecological costs. Since the reform and opening up, China has become a 'superpower' in energy consumption. It has also become a big producer of greenhouse emissions.

As mentioned in earlier chapters, the Twelfth Five-Year Plan timely proposed a green development strategy for China. It was the first Five-Year Plan ever to tackle green development systematically and indicated China's dedication in reversing the long-standing trend toward ecological deterioration. China is expected to lead the world in global green development. In several ways, China can make an effort to optimize its energy structure. First, with the continuous advancement in high technology, non-fossil energy sources, such as solar, wind and other renewable forms of energy can be better utilized. Second, in the consumption of fossil energy, the use of coal can be reduced. Third, also thanks to technological advancement, fossil fuels such as coal can be used in a much cleaner fashion. The share of non-fossil energy use with total energy consumption surpassed 10 % by 2015. According to the environmental objectives proposed by the State Council in 2009, the share of the consumption of renewable energy forms will reach 15 % by the year 2020. Conservatively estimated, the share of non-fossil energy consumption will reach about 20 % by 2020 and about 25 % by 2030.

Y. Jiang, *Green Development in China*,
SpringerBriefs in Economics, DOI 10.1007/978-981-10-0693-7_9

China is expected to become a world's leader in developing a low-carbon economy. Published sources by various authorities predicted that the average annual growth of demand for natural gas is going to reach 1.4 % over the period of 2008–2035. China, however, was predicted to have the fastest growth rate of 6 % and account for over one quarter of the overall growth in global demand over the period. Innovations in low-carbon technologies are propelled by China's huge domestic market and investment demand. According the estimates from various sources, China will account for about 20, 25, 30 and 20 % of the world's total innovations of low-carbon technologies respectively in the fields of solar energy, wind energy, nuclear energy, and hybrid cars.

We also expect China to become the fastest growing country in forestry resources, rapidly closing on the gap with the most forested countries in the world. China can hopefully become the nation with the largest manmade forest carbon storage and the largest manmade forest carbon sinks. China will also raise its water conservation efforts significantly, making an effort in striving to tackle desertification, promote sand fixation, and soil and water conservation, protect natural resources, and maintain the biological diversity.

We hope that the environmental quality can be steadily improved as China moves on through its path of environmentally sustainable development. Pollution emissions will be brought down gradually to below the environmental self-purification capacity. China shall make an unswerving effort in coordinating the production and consumption between human activities and the natural ecosystem, minimizing the environmental costs of its economic and social development. Overall, China is expected to make initiative in giving a positive and responsible response to the challenges the world is currently facing with respect to the prevalent environmental and ecological problems, and in doing so, make its utmost contribution to the joint green development of the future world.

MIX
Papier aus verantwortungsvollen Quellen
Paper from responsible sources
FSC® C105338

If you have any concerns about our products,
you can contact us on
ProductSafety@springernature.com

In case Publisher is established outside the EU,
the EU authorized representative is:
Springer Nature Customer Service Center GmbH
Europaplatz 3, 69115 Heidelberg, Germany

Printed by Libri Plureos GmbH
in Hamburg, Germany